*Rivers
of the
United States*

Rivers
of the
United States

VOLUME V
PART B: THE GULF OF MEXICO

Ruth Patrick

FRANCIS BOYER CHAIR OF LIMNOLOGY
THE ACADEMY OF NATURAL SCIENCES OF PHILADELPHIA

WILEY

JOHN WILEY & SONS, INC.

Published by John Wiley & Sons, Inc., Hoboken, New Jersey
Published simultaneously in Canada

For general information on our other products and services or for technical support, please contact our
Customer Care Department within the United States at (800) 762-2974, outside the United States at
(317) 572-3993 or fax (317) 572-4002.

Wiley also publishes its books in a variety of electronic formats. Some content that appears in print may
not be available in electronic books. For more information about Wiley products, visit our web site at
www.wiley.com.

Library of Congress Cataloging-in-Publication Data:

Patrick, Ruth.
    Rivers of the United States.

    1. Rivers—United States.   I. Title.

GB1215.P29   551.48'3'0973   93-27583

ISBN 0-471-30345-3 (v. 1)
ISBN 0-471-10752-2 (v. 2)
ISBN 0-471-30346-1 (v. 3)
ISBN 0-471-19741-6 (v. 4, Pt. A)
ISBN 0-471-19742-4 (v. 4, Pt. B)
ISBN 0-471-30347-X (v. 4, set)
ISBN 0-471-30348-8 (v. 5, Pt. A)
ISBN 0-471-30349-6 (v. 5, Pt. B)

Printed in the United States of America

10 9 8 7 6 5 4 3 2 1

Contents

Preface

Texas is the largest state in the continental United States. It is very diverse as to environment and industrial activities. In the eastern part of the state the rainfall and general climate conditions are very similar to those of the tributaries of the Mississippi River, whereas the western part of the state is arid and the rivers have very high conductivity and are often very hard.

Industrial activity is very different in the two parts of the state. In the eastern part of the state, industrial activities are very diverse. It is in this part of the state that oil was first discovered in the early part of the twentieth century and has dominated a considerable amount of the industrial activity. In the western part of the state, ranches and agriculture are the principal activities. The rivers are dammed and diverted and used heavily for these purposes. As a result, the Rio Grande no longer flows into the Gulf of Mexico, as not enough water is available at its outlet. The chemical characteristics of the river waters are very variable, and the associations of aquatic life indicate these conditions.

For this volume I have chosen the Sabine River and the Guadalupe River to represent the biological and chemical characteristics of rivers in the eastern part of the state. The Rio Grande, which forms the boundary line in Texas between Mexico and the United States, and the Pecos River have been chosen to represent western, arid-region rivers.

Reaches of the Guadalupe and Sabine rivers have been used to represent water conditions in the central and eastern part of Texas. The Guadalupe River is a hard, alkaline stream dominated by magnesium and calcium carbonates and bicarbonate alkalinity. The reaches described are in relatively unpolluted areas of the Guadalupe River. The Guadalupe River flows through the Edwards Plateau, descends the

Balcones Escarpment, and enters the Texas extension of the Coastal Plains Physio-graphic Province.

The Sabine River flows southeast and south for 380 mi from Hunt County in eastern Texas to Sabine Lake, and through the Sabine pass into the Gulf of Mexico. The entire Sabine basin lies within the Gulf Coast Plain Section of the Coastal Plain Physiographic Province ((N. M. Fenneman, *Physiography of the Eastern United States*, McGraw-Hill, New York, 1938). The Sabine River basin lies within three major resource areas: the Black Plain Prairies, East Texas Timberlands, and Coastal Prairie. The upper Sabine is a soft to medium-hard river and the nitrates and phosphates indicate some agricultural activity. However, the water chemistry indicates fairly normal conditions. The lower Sabine has soft to moderately alkaline slightly saline water. It is just above the reach where the intrusion of the tide meets the freshwater stream. Therefore, occasionally salt water intermingles with fresh water.

The aquatic life of the upper Sabine River is typical of a freshwater stream, whereas that which we studied in the lower reach is freshwater but also has species tolerant of a small amount of salinity.

For the Rio Grande and the Pecos, the principal activities of the watershed are agriculture and ranching. As a result, the rivers are diverted by dams and water is extracted for irrigation purposes. Indeed, the Rio Grande at the present time does not flow into the Gulf of Mexico because of this diversion. These are typically hardwater rivers.

The ecosystems for the Sabine and Guadalupe rivers have been developed carefully because of studies made by the Academy of Natural Sciences. Chemists as well as specialists in the various forms of aquatic life were present, and a team approach was taken for the studies. The studies for the Pecos and Rio Grande have been made by scientists from various universities and state agencies. These studies have, in some cases, been general but are specific in their objectives, such as studying the fish populations. These rivers have been very badly damaged by pollution mostly from agricultural sources, and therefore the ecosystems reflect these conditions. I have tried to describe functional ecosystems of both the Rio Grande and Pecos rivers, but they are not as specific as the descriptions of ecosystems in the Guadalupe and Sabine rivers. The descriptions of the ecosystems are for a larger area than they are for the ecosystems described in the Sabine and Guadalupe rivers.

The literature is very scattered as it pertains to the Rio Grande and Pecos rivers. The studies have not been as intense for given areas as they have been in the Sabine and Guadalupe rivers. For these reasons, the ecosystems are not developed as specifically. In the Guadalupe and Sabine rivers the ecosystems have been described for a limited area and the data were gathered by a team of scientists who were specialists in the various groups of organisms. In the western rivers, the Pecos and the Rio Grande, the ecosystems are more general than the detailed studies carried out by the Academy of Natural Sciences for the Sabine and Guadalupe rivers. However, I was able to construct functioning ecosystems by bringing together the data that were available.

Acknowledgments

Many people have been helpful in the preparation and execution of these reports on the Guadalupe, Sabine River, Rio Grande, and Pecos rivers.

In addition to the staff of the Academy of Natural Sciences who did the research and preparation of the data on which this volume is based, I want to thank Dr. Marion Toole of the Fish and Game Commission. I also wish to thank Dr. Fred Hansler, Dr. George W. Cox, and Dr. V. M. Ehler of the Texas Board of Health. They were particularly helpful in the preparation of information concerning the Guadalupe River. Over time, many people have been very helpful in obtaining the research and other reports on the Guadalupe River. They are Mr. William Brondyke of the DuPont Company; Mr. W. H. Shearer, Jr. of the DuPont Company; Dr. Robert J. Kemp, Director of Fisheries, Texas Park and Wildlife Service; Dr. Minter J. Westfall, Jr., University of Florida; and Dr. Henry B. Dodgen, Fish and Game Commission. I also wish to thank Mr. Andrew Nickolaus, who helped us to execute the survey of the Guadalupe River.

For our studies of the Sabine River in Texas, I wish to express my appreciation for their help (at the time of these studies) to Dr. David G. Cooke of the National Museum of Natural Sciences of Ottawa, Ontario, Canada; Dr. Margaret Simpson of the Biology Department of Adelphi University, Garden City, New York; Dr. Victor Zullo, Department of Geology, California Academy of Sciences, San Francisco; Mr. Donald Boesch of the Virginia Institute of Marine Science, Gloucester Point, Virginia; and Dr. Fenner A. Chace, Jr., Dr. Henry Roberts, and Dr. Horton H. Hobbs of the National Museum of Natural History, Smithsonian Institution. With regard to studies of the upper Sabine River I particularly wish to acknowledge Dr. Donald F. Charles of the Academy of Natural Sciences and Dr. Raymond W. Bouchard also of the Academy of Natural Sciences. I wish to thank Mr. Mike Chaffin and Dr. Karolyn Hardaway of Texas Eastman, who assisted us in these studies of the upper Sabine River; also Dr. David Hall, Mr. Dwight Addy, and Mr. Freddy Armstrong.

For their help on the Rio Grande, I wish to thank Dr. Tom Waller and Dr. Jim Kennedy of the Institute for Applied Sciences at the University of North Texas; and Dr. Forrest John, Environmental Protection Agency. For his help on the Pecos River, I wish to thank Dr. Willard J. Gibbons, U.S. Geological Survey.

The most helpful person in providing information concerning these rivers has been Professor John Cairns of Virginia Polytechnic Institute, who at one time worked on our staff and has continued to be most helpful to me in gathering information. He was a leader on the survey teams when the studies were made on the Guadalupe and Sabine rivers, and has facilitated my contacting some of the scientists in Texas. I also particularly want to acknowledge Susan Durdu, who has been my technical assistant and constant help during the preparation of these volumes. She has not only helped with the securing of the literature and the technical editing of the reports but also is responsible for getting together the information to make the maps of the Rio Grande and Pecos Rivers and made the rough drawings of the maps. I also want to thank Su-Ing Yong; who has drawn the maps used in this book.

I also wish to acknowledge the help of the editors of the John Wiley & Sons, Inc., particularly Dr. Philip C. Manor, Jim Harper, Bob Hilbert, and Diana Cisek.

The many scientists of the Academy staff who over time have carried out research on these rivers are very numerous. Besides Dr. John Cairns, I particularly want to recognize Mr. Sam Fuller, now deceased; Robert Grant, Jr.; Jules Loos, formerly of the Academy of Natural Sciences, now with the Potomac Electric Power Company; Dr. Jay Richardson, formerly of the Academy of Natural Sciences; Ms. Fairie Lyn Carter; Dr. Mary Gojdics; Mr. John H. Wallace (deceased); Mr. Charles B. Wurtz (deceased); Dr. Selwyn S. Roback (deceased); Dr. Charles Reimer of the Academy of Natural Sciences; Dr. John T. Gallagher (deceased); Dr. Reeve M. Bailey of the University of Michigan; and Dr. C. L. Smith of the University of Michigan.

The literature for these volumes is widespread, but I want to mention particularly *An Introduction to the Aquatic Insects of North America* by R. W. Merritt and K. W. Cummins (Kendall/Hunt Publishing, Dubuque). The third edition of this book has been especially useful in developing the information concerning the various insects that are mentioned in this volume.

I also want to acknowledge the financial help of many people. The John and Alice Tyler Award, given to me many years ago and invested, has provided the income that has in part, made possible, the writing of this book. I also wish to acknowledge the help of the E.I. DuPont Company, the Phoebe W. Haas Charitable Trust B, and the Academy of Natural Sciences.

CHAPTER 1

Sabine River

===

INTRODUCTION

The Sabine River in the southwestern United States flows southeast and south for 380 mi from Hunt County in eastern Texas to Sabine Lake and through Sabine Pass into the Gulf of Mexico. It has a drainage basin of 9,756 mi^2, which lies in Texas and in the Louisiana Coastal Plain. Discharge of the river varies from 372 to 61,200 ft^3/s. The Texas–Louisiana boundary follows the river, Sabine Lake, and Sabine Pass (Figure 1.1).

The entire Sabine basin lies within the Western Gulf Coastal Plain Section of the Coastal Plain Physiographic Province (Fenneman, 1938). According to the biotic provinces developed by Blair (1950), the upper third of the Sabine lies within the Texan Province, while the remainder lies within the more humid Austroriparian Province. Rolling plains characterize the upper fourth of the basin, but the topography becomes progressively flatter downstream (Hughes and Leifeste, 1965). Within the lower Sabine, the topography grades from low hills in the upper half (i.e., near Ruliff, Texas) to flat and often swampy topography in the lower half (i.e., in Orange County, Texas) (Seidensticker, 1980).

The Sabine River basin lies within three major land resource areas: the Black Land Prairies, East Texas Timberlands, and the Coastal Prairie. The Black Land Prairies land resource area encompasses approximately 960 mi^2 within the basin and comprises the entire area upstream of Iron Bridge Dam and the extreme upper portion of Lake Fork Creek Subwatershed. The East Texas Timberlands area comprises about

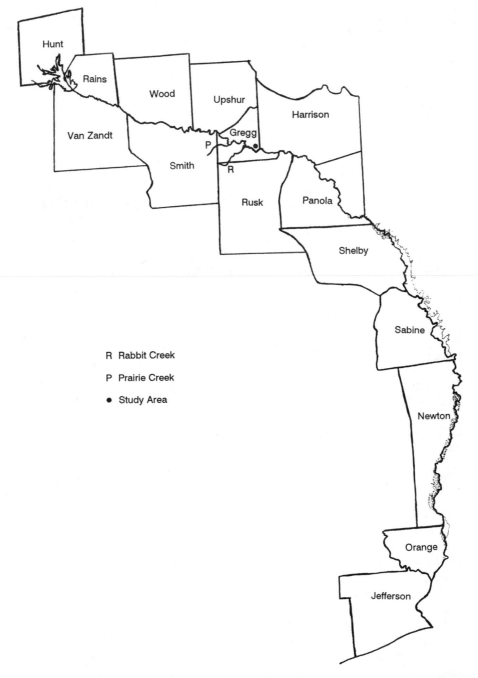

R Rabbit Creek

P Prairie Creek

● Study Area

FIGURE 1.1.   Map of the Sabine River.

88% of the basin. The Coastal Prairie land resource area comprises approximately the lower 10 mi of the Sabine River basin.

PHYSICAL CHARACTERISTICS

Geology

Geologically, the Sabine basin is underlain with sedimentary deposits that range from Upper Cretaceous to Recent in age. The oldest deposits are found in the northwestern part of the basin. Tertiary deposits of sandstones, shales, and unconsolidated sediments are prevalent in most of the basin, while Pleistocene deposits of clay, silt, sand, and gravel characterize much of the lower basin. Recent alluvium underlies and borders most of the middle and lower river (Hughes and Leifeste, 1965).

Watershed

With the exception of Black Land Prairies (predominantly *Andropogon* spp. and *Stipa leucotrichia*) in the northwest, the presettlement watershed of the Sabine River was covered with pine and hardwood forests. A longleaf pine (*Pinus palustris*) covered most of the lower Sabine watershed, but repeated clearing has largely replaced this assemblage with a Southern Mixed Forest characterized by sweetgum (*Liquidambar styraciflua*), postoak (*Quercus stellata*), blackjack oak (*Q. marilandica*), and yellow pine (*P. echinata*) in the uplands. Bottomland forests are characterized by *L. styraciflua,* magnolia (*Magnolia grandiflora*), tupelo gym (*Nyssa sylvatica*), and water oak (*Q. nigra*). Cypress–tupelo (*Taxodium–Nyssa*) swamps are extensive in the flatter, downstream reaches and along some of the tributaries, but these forests grade into coastal grassy wetlands within the estuary [Academy of Natural Sciences of Philadelphia (ANSP), 1954; Blair, 1950; Hubbs, 1957; Sala et al., 1979; Seidensticker, 1980]. Agricultural activity was formerly extensive within the lower basin, but yellow pine and hardwood forestry dominate current land use. Industrial activity is considerable within the estuary (ANSP, 1970; Sala et al., 1979).

The climate within the Sabine basin ranges from subhumid in the northwest to humid or subtropical in the lower Sabine. Precipitation, which occurs almost exclusively as rain, averages 142 cm yr^{-1}. The 30-year mean at Bon Wier indicates that precipitation occurs throughout the year, with minima occurring in September and October. Nonetheless, precipitation can be highly variable, particularly in the upper basin, where drought conditions have resulted in occasional intermittency in the main stem (Hughes and Leifeste, 1965).

River Structure

The drainage area of the upper part of the Sabine River is about 4846 mi^2. The lower basin or state line portion has a contributing area of some 4910 mi^2, of which 2550 mi^2 lies within Texas and 2360 mi^2 lies in Louisiana. The total watershed is approximately 9756 mi^2, of which 76% lies within the boundaries of Texas.

The lower Sabine River is typical of many low-gradient, meandering Coastal Plain rivers. The river flows through a wide, well-developed floodplain with numerous sloughs, overflow channels, and wetlands (Hughes and Leifeste, 1965). The channel is relatively wide. Sala et al. (1979) reported that the river at West Bluff (river km 31) and Ruliff (km 60) was between 75 and 110 ft (23 to 34 m) wide during normal flow. Farther upstream at Bon Wier (km 150), they reported that the river was 30 to 50 ft (9 to 15 m) wide during low flow, but expanded to 100 yards (92 m) during high flow. Depths at Ruliff and West Bluff ranged from 10 to 20 ft (3 to 6 m). Seidensticker (1980) reported that widths increased substantially (e.g., $\frac{1}{4}$ mile; 155 m) as the river neared the estuary and that depths varied considerably (1 to 75 ft; 0.3 to 22.9 m). Sand is the predominant substrate, but mud and hard clay can be found in some shallow areas. Snags, woody debris, and undercut banks are moderately common (ANSP, 1954; Sala et al., 1979; Seidensticker, 1980).

Variability in flow characterized the preimpoundent lower Sabine River. Annual mean discharge near Ruliff ranged from 1760 to 17,210 ft³/s (49.8 to 487.3 m³ s⁻¹) between 1925 and 1962. Instantaneous discharge has been even more variable, ranging from 270 to 121,000 ft³/s (7.6 to 3426.4 m³ s⁻¹) (Hughes and Leifeste, 1965). Since 1966, flow in the lower Sabine has been regulated by Toledo Bend Dam. Nonetheless, flows are still highly variable. For instance, annual mean discharge at Ruliff during water year 1969 (348.9 m³ s⁻¹) was almost triple what it was during water year 1970 (117.0 m³ s⁻¹) [U.S. Geological Survey (USGS), 1975a].

Discharge in the lower Sabine River is significantly influenced by discharge from both the upper basin (via Toledo Bend Reservoir) and local tributaries (via Bayou Anacoco and Big Cow Creek). Discharge maxima in reservoir releases (i.e., near Burkeville) generally correspond to discharge maxima, more than 160 km downstream near Ruliff. On the other hand, tributary inputs within the lower basin tend to moderate peak discharge by increasing discharge minima. During water year 1969, for example, there was relatively little difference between the maximum instantaneous discharge recorded near Ruliff and the one recorded near Burkeville. On the other hand, the minimum discharge near Ruliff was more than an order of magnitude greater than it was just below the reservoir (USGS, 1975a). In addition to its influence on seasonal patterns in discharge, hydroelectric generation at Toledo Bend Dam results in severe daily fluctuations in discharge and river stage. Variations in river stage ranging from 4 to 10 ft (1.2 to 3.0 m) have been recorded in the upper half of the lower Sabine (Seidensticker, 1980).

The lower Sabine River is relatively turbid, largely because of high silt loads incurred in the upper and middle basins. Turbidities at Bon Wier, Ruliff, and West Bluff ranged from 3 to 20 NTU, 10 to 110 NTU, and 7 to 99 NTU, respectively, for six collection dates during 1976 to 1977. Concurrent Secchi disk readings ranged from 15 to 48, 9 to 25, and 13 to 28 in., respectively (Sala et al., 1979). The ranges for JTU turbidities for four collection dates during 1979 ranged from 13 to 32, 15 to 34, and 8 to 46 JTU, respectively (Seidensticker, 1980).

The lower Sabine is a warm-water stream with temperatures commonly ranging from 5 to 32°C (ANSP, 1954, 1970; Sala et al., 1979; Seidensticker, 1980). For example, temperatures at Ruliff ranged from 6.0 to 32.0°C during water year 1970. Temperatures were usually greater than 25.0°C from late May through mid-October. Temperatures less than 10°C were recorded only during January (USGS, 1975b).

CHEMICAL CHARACTERISTICS

Upper Sabine

The study area was just above the boundary between Gregg and Harrison counties on the Sabine River. It was a short distance below the junction of Prairie Creek and Rabbit Creek (Figure 1.1). The hardness ranged between 52 and 52.3 mg L^{-1} at the time it was sampled, which was in October 1995 (Table 1.1). Dissolved oxygen was below saturation, which is probably caused by runoff from agricultural activities and from oil well activities in the watershed, which is also indicated by the relatively high chloride concentrations, which are higher than they would be for a natural stream. The nitrates and phosphates were also higher than one would find in a pristine stream, and these concentrations indicate the agricultural activities that take place in the watershed. The dissolved oxygen is similar in concentration to that found in the lower Sabine. However, there were occasions when supersaturation occurred in the lower Sabine, probably due to excessive algal activity.

TABLE 1.1. *Mean, Minimum, Maximum Values, and Standard Deviations for Water Quality Parameters in Samples Collected October 9–11, 1995, from the Upper Sabine River*

Parameter	Mean	Min.	Max.	Std. Dev.
Temperature (°C)	20.73	20.20	21.00	0.46
pH	7.17	7.10	7.30	0.11
Dissolved oxygen (mg L^{-1})	7.70	7.50	8.00	0.26
Conductivity (µS cm^{-1})	261	245	273	14
TSS (mg L^{-1})	26.00	25.80	26.20	0.20
BOD_5 (mg L^{-1})	1.85	1.55	2.10	0.28
Total hardness (mg L^{-1})	52.31	52.00	52.90	0.51
Calcium hardness (mg L^{-1})	34.93	32.70	36.70	2.04
Alkalinity (mg L^{-1})	36.90	33.30	41.80	4.40
Sulfate (mg L^{-1})	29.65	28.87	31.11	1.27
Chloride (mg L^{-1})	49.17	45.00	55.00	5.20
DOC-C (mg L^{-1})	5.74	5.66	5.86	0.10
$NH_3 + NH_4$-N (mg L^{-1})	0.065	0.012	0.154	0.078
NO_2-N (mg L^{-1})	0.012	0.008	0.020	0.007
$NO_3 + NO_2$-N (mg L^{-1})	1.31	1.05	1.53	0.24
TKN-N (mg L^{-1})	0.830	0.761	0.907	0.073
Orthophosphate-P (mg L^{-1})	0.093	0.072	0.109	0.019
Total phosphate-P (mg L^{-1})	0.191	0.170	0.221	0.027

Source: ANSP (1996).

Lower Sabine

Water in the lower Sabine River is soft, moderately acidic to moderately alkaline, slightly saline, and stained with organic acids (Table 1.2). This water chemistry is the result of interactions of mineral-rich waters originating from upstream reaches and mineral-poor waters entering from local tributaries. Under very high tides, salt water from the estuary may invade the area.

TABLE 1.2.　*Selected Chemical Characteristics of Three Stations in the lower Sabine River, Louisiana and Texas, 1972-1977[a]*

	Bon Wier	Ruliff	West Bluff
Inorganic constituents			
Chloride (mg L^{-1})			
Mean	26	21	19
Range	20–40	10–31	6–33
Sulfate (mg L^{-1})			
Mean	20	16	19
Range	13–38	9–24	8–34
Specific conductance (μS cm^{-1})			
Mean	205	180	165
Range	165–250	145–220	90–335
Alkalinity			
Total alkalinity (mg L^{-1})			
Mean	28	25	25
Range	22–42	11–43	8–61
pH			
Mean	7.0	7.0	7.0
Range	6.9–7.3	6.7–7.4	6.7–7.6
Nutrients			
NO_3-N (mg L^{-1})			
Mean	0.11	0.15	0.08
Range	<0.01–0.21	<0.10–0.37	<0.10–0.26
NH_3-N (mg L^{-1})			
Mean	0.12	0.10	0.08
Range	0.02–0.42	0.04–0.21	0.01–0.34
PO_4-P (mg L^{-1})			
Mean	0.03	0.03	0.03
Range	<0.01–0.06	0.02–0.04	0.02–0.04
Dissolved oxygen			
Concentration (mg L^{-1})			
Mean	8.3	8.0	7.5
Range	6.4–12.4	6.4–12.0	5.9–11.6
Percent saturation			
Mean	89	85	82
Range	78–101	80–95	64–97
Color (Pt-Co units)			
Mean	36	40	53
Range	15–90	26–50	23–120

Source: After Sala et al. (1979).

[a]Data from Bon Wier and Ruliff are from July 1976 through Oct. 1977. Data from West Bluff are from Feb. 1972 through Oct. 1977.

Waters draining the northwestern headwaters are hard to medium-hard. Water hardness is reduced as the river passes through the middle basin, but chloride concentrations remain high as a result of both natural inputs and brine pollution from oil fields. Natural levels of chloride concentrations for tributaries upstream from Toledo Bend Reservoir usually range between 20 and 100 ppm. A major exception occurs in Grand Saline Creek, Van Zant County, where water draining a salt flat contained chloride concentrations as high as 39,200 ppm. Otherwise, levels in excess of 100 ppm were usually the result of oil field brine pollution. Chloride concentrations in excess of 20 ppm persist throughout the lower Sabine as a result of these upstream inputs (Hughes and Leifeste, 1965).

Macronutrient concentrations in the lower Sabine are moderate but probably not limiting to primary production. Presumably, these moderate levels result from the prevalence of heavily forested watersheds combined with relatively few municipal inputs. The differences between nitrate-nitrogen concentrations near Bon Wier and Ruliff evident in Table 1.2 suggest a longitudinal increase in this form of inorganic nitrogen. Significantly, both sources (Sala et al., 1979, and the Academy's survey site) reported that a large part of the organic nitrogen was reduced (i.e., ammonia-nitrogen and/or nitrite-nitrogen rather than nitrate-nitrogen) (Table 1.3).

TABLE 1.3. *Water Chemistry for the Lower Sabine River near Orange, Texas*

	Aug. 1952	Apr. 1953	Aug. 1962	Summer 1969
Inorganic constituents				
Total hardness (ppm)	75	46	77	46
Calcium (ppm)	12	11	13	10
Magnesium (ppm)	11	4	11	4
Sulfate (ppm)	6	25	32	17
Chloride (ppm)	156	46	180	26
Specific conductance				
(μS cm^{-1})	966	197	1440	189
Alkalinity				
Bicarbonate alkalinity	38	30	27	32
(ppm)				
pH	7.3	6.9	7.2	6.9
Nutrients				
NO_3-N (ppm)	0.04	0.09	—	—
NO_2-N (ppm)	0.02	0.003	0.002	0.008
NH_3-N (ppm)	0.22	0.11	0.056	0.413
PO_4-P (ppm)	0.002	0.005	0.042	0.036
SiO_2 (ppm)	12	12	15	9
Dissolved oxygen				
Concentration (ppm)	6.5	6.8	5.9	5.1
Percent saturation[a]	83	76	76	65
Temperature (°C)	31.1	21.5	32.2	30.3

Source: Data for 1952 and 1953 after ANSP (1954); data for 1962 after ANSP (1963); data for 1969 after ANSP (1970). Data for all of these sources are from station 1.

[a]Percent saturation determined from dissolved oxygen concentration and stream temperature. Percentages for 1952, 1962, and 1969 are percent saturations if the stream temperatures were 29.0°C. Actual percent saturations would be somewhat higher.

The lower Sabine River is relatively well oxygenated. Dissolved oxygen concentrations at Bon Wier, Ruliff, and West Bluff ranged between 5.5 and 12.4 mg L^{-1}, while percent saturation ranged between 62 and 104% (Sala et al., 1979; Seidensticker, 1980; USGS, 1975b). Dissolved oxygen concentrations and percent saturations at the Academy's survey site near Orange also suggested relatively oxygenated conditions. Saltwater intrusion at this site during August 1953 and August 1962 resulted in stratification and reduced benthic dissolved oxygen concentrations to less than 1 ppm (ANSP, 1954, 1963.)

Pollution in the lower Sabine River is minimal, with point sources being limited primarily to two paper mills. One of the apparent impacts from these mills is the increase in river color during low flow. Another water quality problem encountered in the lower Sabine concerns elevated concentrations of mercury, the origin of which has not been determined (Sala et al., 1979; Seidensticker, 1980). In contrast to the freshwater portion of the lower Sabine, the estuary receives considerable municipal and industrial pollution (ANSP, 1954, 1963, 1970).

ECOSYSTEM DYNAMICS

Detritus and Dissolved Organic Carbon

Upper Sabine. No measurements have been made of the detrital and dissolved organic carbon concentrations in the upper Sabine. They are probably low, as the river has a moderately high gradient, the stream is very shallow and floods would probably wash out any accumulation of organic debris.

Lower Sabine. Information concerning detrital and dissolved organic carbon (DOC) dynamics in the lower Sabine River is limited. Observations at the Academy's site near Orange indicated that concentrations of suspended detritus were relatively high and that there was some detrital accumulation along the shores (ANSP, 1954). Six sets of total organic carbon (TOC) concentrations were measured between July 1976 and October 1977 by Sala et al. (1979). TOC concentrations were relatively high, ranging from 9 to 18 mg L^{-1} at Bon Wier, 6 to 18 mg L^{-1} at Ruliff, and 7 to 19 mg L^{-1} at West Bluff. No consistent relationship between TOC and either season or discharge was evident.

The organic matter dynamics of the lower Sabine may have many characteristics in common with the Ogeechee River, a subtropical Coastal Plain river located in Georgia. Organic matter dynamics in the Ogeechee are characterized by high allochthonous inputs of DOC, leaf litter, and allochthonous bacteria during floodplain inundation. In contrast, autochthonous production is largely limited to low-flow periods during summer, when the river is shallower and less darkly stained by DOC (Benke and Meyer, 1988). Both the Ogeechee and the lower Sabine have well-developed and densely forested floodplains, subtropical climates, and seasonal floodplain inundation. However, floodplain inundation in the lower Sabine is probably not as predictable nor as extensive as it is in the Ogeechee. In addition, the lower Sabine has higher silt loads and its flow is largely regulated.

While seasonal pulses of organic matter were not evident from the TOC concentrations reported by Sala et al. (1979), more extensive monitoring of the river's color indicates that DOC inputs are associated with floodplain inundation. During water year 1970, for instance, color in the lower Sabine at Bon Wier ranged from 5 to 220 Pt-Co units, with values typically exceeding 100 Pt-Co units during winter and usually less than 50 Pt-Co units from April through September. Similar seasonal patterns occurred near Ruliff, where color ranged fro 15 to 220 Pt-Co units (USGS, 1975b).

Primary Production

Upper Sabine. No measurements were made of the amount of primary production in the upper Sabine. However, since there is a well-developed algal flora in this reach of the river, it is very probable that primary production is relatively high.

Lower Sabine. Chlorophyll *a* concentrations at Bon Wier, Ruliff, and West Bluff averaged 0.003, 0.004, and 0.004 mg L^{-1}, respectively, from July 1976 through October 1977. Concentration maxima were most often recorded during summer, but were also recorded in winter and spring. Algal blooms are apparently infrequent (Sala et al., 1979).

Information on algal distribution and taxonomic composition in the lower Sabine River is largely restricted to investigations at the Academy's survey site near Orange. Algae were found on a variety of habitats at up to a depth of 1 m during a low-water, high salinity survey conducted in August 1952. The flora was characterized by *Spirogyra*, as well as by a variety of blue-greens and diatoms. In contrast, high turbidity during a high water survey conducted the following April (1953) largely limited algae to snags and shallow bank habitats. Blue-greens and diatoms were both moderately common (ANSP, 1954). Saline and moderately turbid conditions also occurred during an August 1962 survey. Algal growth was moderately heavy, although most algae were limited to shallow or snag habitats. The 1962 flora was similar to that recorded during August 1952 (ANSP, 1963).

Low salinity and high turbidity characterized conditions during the summer 1969 survey. Algal growth was extensive but was again largely restricted to either woody or shallow habitats. Blue-greens were abundant and diatoms were common, but green and red algae were scarce. *Microcoleus vaginatus, Oscillatoria princeps, O. retzii, Schizothrix calcicola*, and *S. tenerrima* were the more abundant blue-green algae. Diatoms were especially diverse, with 111 species being recorded during this survey. Of these, *Nitzschia filiformis* and *N. frustulum* var. *perminuta* were the most frequent (ANSP, 1970).

Characteristics of the Macroinvertebrate Fauna

Upper Sabine. The macroinvertebrate fauna was diverse in the upper Sabine, as all the major groups that one would expect to find in fresh water were present. The area was particularly rich in unionid molluscs. Nine taxons belonging to the family Unionidae were present. Also present was the fingernail claim, *Sphaerium striatinum*. The occurrence of *Corbicula fluminea* was discouraging, as this unionid is known to develop large colonies rather rapidly and displace native clams, which now seem diverse and fairly abundant in the river. Several species of decapods were found in the river. A total of three taxons were identified. The insect fauna were diverse in this area. A total of 35 taxons were collected, which is somewhat less than the number collected (42) in 1987 and 41 in 1982. The main differences seem to be fewer

Heteroptera and Diptera. Mayflies and stoneflies, which are indicators of normal water conditions, were similar in the various studies. The fish fauna was very diverse in this area; a total of 35 taxons were collected. The species were diversified and represented various feeding groups. It is quite evident from examining the structure of the communities of various groups of organisms in the river in this area that the stream has a normal diversity and appears to be in a healthy condition.

Lower Sabine. Macroinvertebrate assemblages at the Academy's survey site near Orange were characterized by relatively few freshwater taxons, presumably because of periodically high salinities. *Stenonema* was the most abundant insect during the April 1953 survey, but it was not collected during other surveys. Two amphipods, *Gammarus mucronatus* and *Hyalella azteca*, were abundant among cypress roots during the summer 1969 survey, and a bryozoan (*Plumatella repens*) was abundant on wood. On the other hand, these assemblages included several abundant estuarine taxa. *Rangia cuneata* was the dominant clam during the 1953 survey. Blue crab (*Callinectes sapidus*), white shrimp (*Penaeus setiferus*), fiddler crab (*Uca rapax*), and grass shrimp (*Palaemonetes pugio*) were abundant. A wood-boring clam (*Congeria leucophaeata*), a barnacle (*Balanus*), and a wood-boring isopod (*Sphaeroma destructor*) were abundant on wood (ANSP, 1954, 1970).

Fish Fauna

Upper Sabine. The fish fauna in the upper Sabine was not only diverse but well established. For example, there was an average of 608 bullhead minnows (*Pimephales vigilax*) in 100 m^2. There was also an average of 238 red shiners (*Cyprinella lutrensis*) per 100 m^2. These were the most common fish in this area. The fish collections were separated according to zones, that is, gradual slopes versus steep slopes. A general statement that one could make concerning the number of fish collected of different types at different slopes in the river is that some fish, such as *Cyprinella lutrensis*, were most numerous in zone 1 in the gradual slope, whereas in the moderate slope there were only 58.15 individuals collected in 6.2 m of net (0.32-cm mesh). However, for *Lythurus fumeus* there were 55.64 fish collected in moderate slopes and only 2.01 collected in gradual slopes. Thus it is evident that the slope affects individual species, not the entire fauna. The average density of fish (i.e., number per 25 m^2 in backpack samplers) from different riffle types and raceways indicates that raceways have the greatest number of fish collected and that the mixed cobble and cobble gravel were quite similar in numbers of fish collected.

Lower Sabine. The lower Sabine River supports a significant sport fishery for blue catfish (*Ictalurus furcatus*), channel catfish (*I. punctatus*), flathead catfish (*Pylodictis olivaris*), largemouth bass (*Micropterus salmoides*), and crappie (*Pomoxis* spp.). In addition, estuarine species such as seatrout (*Cynoscion* spp.) and southern flounder (*Paralichthys lethostigma*) enter the fishery in downstream reaches (Seidensticker, 1980).

A series of gill net collections near Bon Wier, Ruliff, and West Bluff yielded a mean catch per unit effort (CPUE) of 12.52 fish per 200 ft. of net and a mean mass per unit effort (MPUE) of 5.80 kg per net. An estuarine/marine species, sand seatrout (*Cynoscion arenarius*), was the most numerous species (24.8% of CPUE), but it accounted for only 1% of the MPUE. Spotted gar (*Lepisosteus oculatus*), longnose gar (*L. osseus*), alligator gar (*L. spathula*), bowfin (*Amia calva*), ladyfish (*Elops saurus*), gizzard shad (*Dorosoma cepedianum*), threadfin shad (*D. petenense*), river carp-

TABLE 1.4.   *Catch per Unit Effort (CPUE; number per 200 ft of net) and Mass per Unit Effort (MPUE; kg per 200 ft of net) for Gill Net Collections in the Lower Sabine River, Texas and Louisiana, June–September 1979*

	CPUE	MPUE
Lepisosteidae		
Lepisosteus oculatus	1.65	1.55
L. osseus	0.35	0.58
L. spathula	0.02	0.67
Amiidae		
Amia calva	0.13	0.34
Elopidae		
Elops saurus	0.65	0.11
Clupeidae		
Brevoortia gunteri	0.02	<0.01
Dorosoma cepedianum	1.02	0.29
D. petenense	.85	0.01
Catostomidae		
Carpiodes carpio	0.57	0.28
Ictiobus bubalus	0.28	0.30
Other Catostomidae	0.17	0.06
Cyprinidae		
Cyprinus carpio	0.09	0.30
Notropis venustus	0.013	<0.01
Ictaluridae		
Ictalurus furcatus	0.11	0.13
I. punctatus	0.46	0.34
Pilodictis olivaris	0.02	0.03
Centrarchidae		
Lepomis spp.	0.52	0.02
Micropterus punctulatus	0.13	0.02
M. salmoides	0.22	0.10
Pomoxis annularis	0.70	0.10
P. nigromaculatus	0.19	0.04
Mugilidae		
Mugil cephalus	0.30	0.11
Sciaenidae		
Aplodinotus grunniens	0.09	0.16
Cynoscion arenarius	3.11	0.06
Micropogonais undulatus	0.28	0.01
Other		
Total	0.47	0.16
Total CPUE and MPUE	12.52	5.80

Source: Modified from Seidensticker (1980).

sucker (*Carpiodes carpio*), smallmouth buffalo (*Ictiobus bubalus*), carp (*Cyprinus carpio*) and *Ictalurus punctatus* were abundant in terms of numbers and/or biomass (Seidensticker, 1980) (Table 1.4).

Boat electrofishing collections in the lower Sabine yielded a mean CPUE of 23.89 fish per 15 min of electrofishing and a mean MPUE of 6.34 kg per 15 min. Another primarily estuarine/marine fish, striped mullet (*Mugil cephalus*), accounted for 28.7% of the mean CPUE and 28.2% of the MPUE. *Lepisosteus oculatus, Amia calva*, finescale menhaden (*Brevoortia gunteri*), *Dorosoma cepedianum, Cyprinus carpio*, bluegill (*Lepomis macrochirus*), redear sunfish (*Lepomis microlophus*), *Micropterus salmoides*, and black crappie (*Pomoxis nigromaculatus*) were common to abundant (Table 1.5). Seining conditions in the main stem were generally unfavorable, but relatively high catches of blacktail shiner (*Notropis venustus*) in both the gill nets and with electrofishing suggest that this species is among the most numerous of the small species. *N. venustus*, ironcolored shiner (*N. chalybaeus*), ribbon shiner (*N. fumeus*), and redfin shiner (*N. umbratilis*) were numerous in seine collections from several lower Sabine tributaries (Seidensticker, 1980). Of these, only *N. venustus* and *N. umbratilis* are characteristic of main-stem rivers. Distribution and habitat preference data suggest that four other cyprinids, central silvery minnow (*Hybognathus nuchalis*), pallid shiner (*Notropis amnis*), emerald shiner (*N. atherinoides*), and ghost shiner (*N. buchanani*), may also be numerous in the main stem (Lee et al., 1980). Three of these, *H. nuchalis, N. amnis*, and *N. atherinoides*, were collected in the main stem either at the Academy's survey site or at downstream estuarine sites (ANSP, 1954).

Rotenone collections in a small bayou adjoining the Academy's survey site near Orange yielded 31 fish taxons during the April 1953 survey. *Mugil cephalus* was the most abundant of these, while red shiner (*Notropis lutrensis*) was common (ANSP, 1954).

COMMUNITIES OF AQUATIC LIFE

These studies of riverine systems support the theory of Eugene P. and Howard T. Odum that in each stable community there are four functional stages of nutrient and energy transfer; (1) detritus producers and primary producers; (2) detritivores and herbivores; (3) omnivores; and (4) carnivores.

Functional Relationships

Upper Sabine. The diversity of species in the upper Sabine station indicates that at each stage of nutrient and energy transfer there were many species, representing several different phylogenetic groups. This structure ensures the transfer of nutrients through the food web because if conditions change, eliminating one group, individuals of other species are present to secure the passage of energy and nutrients through the food web. The species diversity characteristic of natural small streams of this type was excellent.

Lower Sabine. Trophic dynamics in the lower Sabine probably depend primarily on allochthonous inputs from inundated floodplains and secondarily on autochthonous primary production and inputs from Toledo Bend Reservoir. Qualitative

TABLE 1.5. *Catch per Unit Effort (CPUE; number per 15 min) and Mass per Unit Effort (MPUE; kg per 15 min) for Boat Electrofishing in the lower Sabine River, Texas and Louisiana, August–October 1979*

	CPUE	MPUE
Lepisosteidae		
Lepisosteus oculatus	1.68	0.60
L. osseus	0.04	0.03
Amiidae		
Amia calva	0.71	1.12
Clupeidae		
Brevoortia gunteri	0.82	0.01
Dorosoma cepedianum	2.61	0.15
Catostomidae		
Minytrema melanops	0.18	0.10
Moxostoma poecilurum	0.25	0.07
Other Catostomidae	0.19	0.07
Cyprinidae		
Cyprinus carpio	0.29	1.71
Notropis venustus	0.21	<0.01
Ictaluridae		
Total Ictaluridae	0.22	0.08
Centrarchidae		
Lepomis gulosus	0.46	0.02
L. macrochirus	3.36	0.05
L. megalotis	0.71	0.05
L. microlophus	1.57	0.05
Micropterus punctulatus	0.46	0.10
M. salmoides	1.93	1.11
Pomoxis nigromaculatus	0.50	0.07
Other Centrarchidae	0.54	0.04
Mugilidae		
Mugil cephalus	6.82	1.79
Percithyidae		
Morone mississippiensis	0.21	0.03
Other		
Total	0.16	0.06
Total CPUE and MPUE	23.89	6.34

Source: Modified from Seidensticker (1980).

macroinvertebrate collections suggest an abundance of benthic detritivores, including oligochaetes, *Hexagenia*, and chironomids. They also suggest an abundance of filter-feeding bivalves, including Sphaeriidae and *Corbicula fluminea*. Both of these groups probably feed heavily on fine detritus and free-living bacteria.

Fish collections in the lower Sabine were dominated by benthic feeding omnivores (e.g., *Carpiodes carpio, Ictiobus bubalus,* and *Cyprinus carpio*), generalist omnivores (e.g., *Mugil cephalus*), planktivores (e.g., *Dorosoma* spp., *Brevoortia gunteri,* and immature centrarchids), and medium-sized to large predators (e.g., *Lepisosteus* spp., *Amia calva,* and *Micropterus salmoides*).

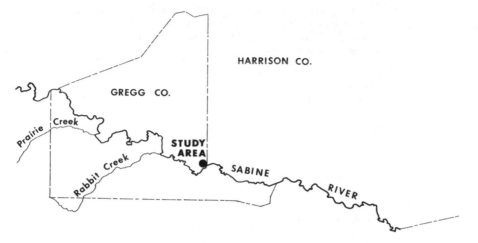

FIGURE 1.2.   Map of the study site on the Sabine River.

Structure of Aquatic Ecosystems

The Sabine River has been studied over many years by the Academy of Natural Sciences. Therefore, ecosystems in the upper area as well as near its mouth are described.

Upper Sabine.   The Sabine River in the study area (Figure 1.2) is a moderate-sized headwater with a typical riffle-pool sequence. The bed of the river consists of rocks and gravel in the riffles and a great deal of sand and mud in the lentic areas.

Vegetative Habitats.   The banks are sandy and there are patches of various aquatic plants found along the edges of the stream and edges of the pools. The main plants found in these vegetative habitats were *Amaranthus tamariscina, Ludwigia decurrens, Polygonum persicaria, Lindernia dubia, Mollugo verticillata,* and *Eclipta alba.* All of these plants grow partially emerged and partially submerged. They typically occur in lentic habitats or along the margins of the stream (Table 1.6).

The blue-green alga, *Anabaena oscillariodes,* was found associated with these plants, as were the desmids *Closterium acerosum* and an unidentified species of *Spirogyra.* Several diatoms were found either growing attached to the plants or among the debris: *Cyclotella comta, Melosira ambigua* found attached to some of the leaves of the plants, and *Skeletonema potamos* in and among the debris. Other algae found attached to the debris and in some cases to the leaves of the plants were *Synedra rumpens* var. *familiaris, S. ulna,* and *S. ulna* var. *contracta.* In and among the debris was *Eunotia exigua.* Attached to the leaves and stems of the plants were *Achnanthes exigua* and *Cocconeis fluviatilis.* Other diatoms found attached to the stems of the plants or the debris were *Amphora ovalis, Cymbella minuta,* and *Gomphonema clevei. G. intricatum* was also in this habitat, as were *Capartogramma crucicula* and *Diploneis puella.* In and among the debris were found *Navicula arvensis, N. canalis, N. capitata, and N. cohnii.* Attached to the stems of some of the plants was *N. confervacea.* Other naviculoid species found in these habitats were *N. integra,*

(text continues on page 24)

TABLE 1.6.   *Species List: Upper Sabine River*

Taxon[a]	V	Lentic		Lotic	
		M	S	G	R
SUPERKINGDOM PROKARYOTAE					
KINGDOM MONERA					
Division Cyanophycota					
Class Cyanophyceae					
Order Chroococcales					
Family Chroococcaceae					
* *Agmenellum quadruplicatum*		X			
Order Nostocales					
Family Nostocaceae					
* *Anabaena oscillariodes*	X	X			
Family Oscillatoriaceae					
* *Microcoleus vaginatus*					X
* *Schizothrix arenaria*		X			
* *S. calcicola*		X			X
* *S. friesii*		X			
SUPERKINGDOM EUKARYOTAE					
KINGDOM PLANTAE					
SUBKINGDOM THALLOBIONTA					
Division Chlorophycota					
Class Chlorophyceae					
Order Chlorococcales					
Family Oocystaceae					
Micractinium sp.					
Family Scenedesmaceae					
* *Scenedesus ecornis*		X			
** *S. quadricauda*		X			
Order Cladophorales					
Family Cladophoraceae					
* *Cladophora glomerata*				X	X
Order Zygnematales					
Family Desmidaceae					
* *Closterium acerosum*	X	X			
* *C. venus*		X			
Family Zygnemataceae					
* *Spirogyra* sp.	X	X			
Division Chromophycota					
Class Bacillariophyceae					
Order Eupodiscales					
Family Coscinodiscaceae					
* *Cyclotella aliquantula*		X			
* *C. atomus*		X			
* *C. comta*	X	X			
* *C. meneghiniana*		X		X	X
* *C. pseudostelligera*		X			
C. stelligera					
* *Melosira ambigua*	X			X	X

(*continued*)

TABLE 1.6.   (*Continued*)

Taxon[a]	V	Lentic		Lotic	
		M	S	G	R
Family Coscinodiscaceae (*cont.*)					
* *M. distans* v. *alpigena*		X		X	
* *M. varians*				X	X
* *Skeletonema potamos*	X				
* *Stephanodiscus astraea* v. *minutula*			X		
* *S. hantzschii*			X		
* *S. invisitatus*			X		
* *S. minutus*			X	X	
Thalassiosira fluviatilis					
T. visurgis					
Order Biddulphiales					
Family Biddulphiaceae					
* *Biddulphia laevis*		X			
Order Fragilariales					
Family Fragilariaceae					
* *Synedra rumpens* v. *familiaris*	X			X	
* *S. ulna*	X	X			
* *S. ulna* v. *contracta*	X				
Order Eunotiales					
Family Eunotiaceae					
* *Eunotia exigua*	X	X		X	
* *E. major*		X			
* *E. monodon*		X			
Order Achnanthales					
Family Achnanthaceae					
* *Achnanthes exigua*	X	X		X	
* *A. minutissima*		X		X	
* *Cocconeis fluviatilis*	X				X
Order Naviculales					
Family Cymbellaceae					
* *Amphora ovalis*	X			X	
* *A. submontana*		X		X	
* *Cymbella minuta*	X	X			X
* *C. minuta* v. *silesiaca*		X			
* *C. naviculiformis*		X			
Family Gomphonemaceae					
* *Gomphonema clevei*	X	X		X	
* *G. grunowii*		X		X	
* *G. intricatum*	X			X	
* *G. parvulum*		X			
Family Naviculaceae					
* *Anomoeneis folis*		X	X		
* *Caloneis bacillum*		X	X		
* *Capartogramma crucicula*	X				
* *Diploneis puella*	X				
* *Frustulia rhomboides* v. *crassinervia*		X	X		
* *F. weinholdii*		X	X		
* *Gyrosigma nodiferum*		X	X		

TABLE 1.6. (*Continued*)

Taxon[a]	V	Lentic		Lotic	
		M	S	G	R
Family Naviculaceae (*cont.*)					
* *G. obscurum*		X	X		
* *G. spencerii*		X	X		
* *Navicula arvensis*	X	X	X		
* *N. atomus*		X	X		
* *N. biconica*		X	X		
* *N. canalis*	X	X	X		
* *N. capitata*	X	X	X		
* *N. cohnii*	X				
* *N. confervacea*	X				
* *N. contenta* v. *biceps*		X	X	X	
* *N. cryptocephala*		X	X		
* *N. grimmei*		X	X		
* *N. integra*	X	X			
* *N. minima*	X	X	X		
* *N. mutica*		X	X		
* *N. ochridana*		X	X		
* *N. omissa*		X	X		
* *N. paratunkae*	X	X	X		
* *N. paucivistata*		X	X		
* *N. pelliculosa*	X			X	
* *N. placentula*		X	X		
* *N. pupula*		X	X	X	
N. pupula v. *mutata*		X	X	X	
* *N. pupula* v. *rectangularis*	X			X	
* *N. rhynchocephala* v. *germainii*	X				
* *N. salinicola*	X				
* *N. symmetrica*	X				
* *N. tenera*		X	X		
* *N. texana*	X				
N. tripunctata v. *schizonemoides*				X	
* *Neidium affine* v. *longiceps*		X	X		
* *Pinnularia biceps*	X			X	
* *P. intermedia*	X		X		
* *P. major*	X		X		
* *P. obscura*			X		
* *P. subcapitata*			X		
Order Bacillariales					
Family Nitzschiaceae					
* *Bacillaria paradoxa*	X	X	X	X	
* *Denticulata elegans*	X				
* *Nitzschia accomodata*				X	
* *N. acicularis*	X				
* *N. amphibia*	X	X			
* *N. apiculata*				X	
* *N. bacata*				X	

(*continued*)

TABLE 1.6.   (*Continued*)

		Substrate[b]			
		Lentic		Lotic	
Taxon[a]	V	M	S	G	R
Family Nitzschiaceae (*cont.*)					
* *N. biacrula*	X				
* *N. circumsuta*	X				
* *N. clausii*	X	X	X		
* *N. confinis*		X	X		
* *N. constricta* v. *subconstricta*				X	
* *N. filiformis*		X	X		
* *N. frequens*		X	X		
* *N. frustulum*	X				
* *N. frustulum* v. *perminuta*	X	X	X		
* *N. frustulum* v. *subsalina*		X			
* *N. gracilis*				X	
* *N. granulata*		X			
* *N. hungarica*		X			
* *N. intermedia*		X			
* *N. kuetzingiana*				X	
* *N. mediastalsis*	X				
* *N. obtusa* v. *scalpelliformis*	X				
* *N. obtusa*	X				
* *N. palea*		X			
N. reversa		X	X		
N. sigma		X	X		
N. sigmoidea		X	X		
* *N. subtilis*		X	X		
* *N. tarda*		X		X	
N. tryblionella v. *debilis*		X			
N. tryblionella v. *levidensis*		X			
N. tryblionella v. *victoriae*		X			
* *N. umbilicata*	X				
N. vasta		X			
Order Surirellales					
Family Surirellaceae					
Surirella augusta		X	X		
S. brightwellii		X	X		
* *S. salina*	X	X	X		
S. stalagma		X	X		
S. suecica		X	X		
S. tenera v. *nervosa*		X	X		
Division Rhodophycota					
Class Rhodophyceae					
Order Acrochaetinales					
Family Audouinellaceae					
* *Audouinella violacea*					X
SUBKINGDOM EMBRYOBIONTA					
Division Magnoliophyta					
Class Dicotyledoneae					
Order Caryophyllales					

TABLE 1.6.   (*Continued*)

Taxon[a]		Substrate[b]				
			Lentic		Lotic	
	V	M	S		G	R
Family Aizoaceae						
* *Mollugo verticillata*	X	X	X			
Family Amaranthaceae						
* *Amaranthus tamariscina*	X	X	X			
Order Myrtales						
Family Onagraceae						
* *Ludwigia decurrens*	X					
Order Polygonales						
Family Polygonaceae						
* *Polygonum persicaria*	X					
Order Scrophulariales						
Family Scrophulariaceae						
* *Lindernia dubia*	X					
Order Asterales						
Family Compositae						
* *Eclipta alba*	X	X				
Class Monocotyledoneae						
Order Arales						
Family Lemnaceae						
* *Lemna* sp.		X	X			
Order Cyperales						
Family Cyperaceae						
* *Fimbristylis vahlii*	X	X				
KINGDOM ANIMALIA						
SUBKINGDOM PROTOZOA						
Class Mastigophora						
Order Euglenoidida						
Family Euglenidae						
* *Euglena* sp.	X	X	X		X	
*o *Trachelomonas hispida*		X	X			
SUBKINGDOM PARAZOA						
Phylum Porifera						
Class Demospongiae						
Order Haplosclerida						
Family Spongillidae						
unident. sp.	X					
SUBKINGDOM EUMETAZOA						
Phylum Mollusca						
Class Gastropoda						
Order Basommatophora						
Family Ancylidae						
*do *Ferrisia rivularis*	X	X				
Class Bivalvia						
Order Unionoida						

(*continued*)

TABLE 1.6. (*Continued*)

Taxon[a]		Substrate[b]			
		Lentic		Lotic	
	V	M	S	G	R
Family Unionidae					
*dh *Amblema plicata*		X	X	X	
*dh *Arcidens confragosus*		X	X		
* *Fusconaia lananensis*				X	
* *Lampsilis satura*				X	
* *L. teres*				X	
*oh *Megalonaias nervosa*		X	X		
Plectomerus dombeyanus					
*oh *Quadrula pustulosa* v. *mortoni*				X	
*oh *Q. quadrula*			X	X	
*oh *Strophitus undulatus*				X	
*oh *Tritogonia verrucosa*		X	X	X	
*oh *Truncilla donaciformis*			X	X	
*oh *T. truncata*			X	X	
*oh *Utterbackia imbecillis*		X	X		
Order Veneroida					
Family Corbiculidae					
*do *Corbicula fluminea*			X	X	
Family Sphaeriidae					
*do *Sphaerium striatinum*		X	X		
Phylum Annelida					
Class Oligochaeta					
Order Tubificida					
Family Tubificidae					
*do *Branchiura sowerbyi*		X			
Phylum Arthropoda					
Class Crustacea					
Subclass Malacostraca					
Order Decapoda					
Family Cambridae					
*o *Orconectes palmeri* v. *longimanus*		X			X
* *Procambarus dupratzi*		X			X
Family Palaemonidae					
*ho *Palaemonetes kadiakensis*	X			X	
Class Insecta					
Order Odonata					
Suborder Zygoptera					
Family Coenagrionidae					
*c *Argia* sp.		X	X		X
Suborder Anisoptera					
Family Aeshnidae					
*c *Nasiaeschna pentacantha*		X	X		X
Family Corduliidae					
*c *Neurocordulia molesta*					X
Family Macromiidae					
*c *Macromia* sp.		X	X		X

TABLE 1.6. (*Continued*)

Taxon[a]		Substrate[b]				
			Lentic		Lotic	
	V	M	S	G	R
Order Ephemeroptera					
Suborder Schistonota					
Family Baetidae					
*do *Baetis* sp.	X				X
Family Ephemeridae					
*o *Hexagenia* sp.		X	X		
Family Heptageniidae					
*o *Stenacron interpunctatum*					X
*o *S. exiguum*					X
Family Oligoneuriidae					
*o *Isonychia sicca*	X				X
Family Tricorythidae					
*o *Tricorythodes* sp.	X	X	X		
Order Plecoptera					
Family Perlidae					
* *Neoperla* prob. *clymene*				X	
Order Hemiptera					
Family Belostomatidae					
*c *Belostoma lutarium*	X	X	X		
Family Nepidae					
*c *Ranatra* sp.	X	X	X		
Family Veliidae					
* *Rhagovelia* sp.	X			X	X
Order Megaloptera					
Family Corydalidae					
*c *Corydalus cornutus*	X	X	X		
Order Coleoptera					
Suborder Adephaga					
Family Dytiscidae					
*c *Deronectes* spp.		X			
Family Gyrinidae					
*c *Dineutes* spp.	X	X			
*c *Gyrinus* sp.	X				
Family Noteridae					
Hydrocanthus spp.	X				
Suborder Polyphaga					
Family Elmidae					
*o *Stenelmis* spp.		X	X		
Order Trichoptera					
Family Hydropsychidae					
*o *Cheumatopsyche* spp.				X	X
*o *Hydropsyche simulans*				X	X
*o *H. orris*				X	X
*o *Macrostenum carolina*				X	X
*o *Potomyia flava*				X	X

(*continued*)

TABLE 1.6.   (*Continued*)

Taxon[a]	V	Lentic		Lotic	
		M	S	G	R
Family Hydroptilidae					
*h *Hydroptila* spp.	X				
Family Leptoceridae					
*o *Nectopsyche candida*	X				
Family Polycentropodidae					
*o *Cyrnellus* spp.	X				
Order Diptera					
Suborder Nematocera					
Family Chironomidae					
*o *Ablabesmyia rhamphe*	X	X		X	
*o *Ablabesmyia* sp.	X	X		X	
*h *Glyptotendipes* sp.	X	X			
*o *Polypedilum* spp.	X				
*o *Rheotanytarsus* sp.				X	
*o *Thienemannimyia* sp.	X	X		X	
Family Simuliidae					
* *Simulium* spp.				X	
Phylum Bryozoa					
Class Ectoprocta					
Order Phylactolaemata					
Family Plumatellidae					
*o *Plumatella emarginata*					X
*o *P. repens*					X
Phylum Chordata					
Subphylum Vertebrata					
Class Osteichthyes					
Order Lepisosteiformes					
Family Lepisosteidae					
*c *Lepisosteus oculatus*	X	X	X		
*c *L. osseus*	X	X	X	X	
Order Clupeiformes					
Family Clupeidae					
*o *Dorosoma cepedianum*	X	X	X	X	X
Order Cypriniformes					
Family Catostomidae					
*o *Carpiodes carpio*			X	X	X
*c *Cyprinella lutrensis*			X		X
*c *C. venusta*			X		
*c *Lythurus fumeus*			X	X	
*c *Macrhybopsis aestivalis*		X	X	X	X
Family Cyprinidae					
*o *Hybognathus nuchalis*	X	X	X		
* *Notropis amnis*		X	X		
*c *N. atrocaudalis*	X	X	X		

TABLE 1.6. (*Continued*)

Taxon[a]	V	Lentic		Lotic	
		M	S	G	R
Family Cyprinidae (*cont.*)					
*c N. buchanani			X		X
* N. sabinae			X		
*o Pimephales vigilax		X	X		
Order Siluriformes					
Family Ictaluridae					
*c Ictalurus furcatus		X	X	X	X
*o I. punctatus		X	X	X	X
*c Noturus nocturnus			X	X	X
*c Pylodictus olivaris		X	X		
Order Cyprinodontiformes					
Family Cyprinodontidae					
*o Fundulus notatus				X	X
Family Poeciliidae					
*o Gambusia affinis	X	X	X		
Order Atheriniformes					
Family Atherinidae					
*c Labidesthes sicculus		X	X	X	
Order Perciformes					
Family Centrarchidae					
*c Lepomis humilis	X	X	X		
*o L. macrochirus	X	X	X		
*c L. megalotis	X	X	X	X	X
*o Micropterus punctulatus		X	X	X	
*c Pomoxis annularis	X	X	X		
Family Percidae					
*c Ammocryptia vivax			X		X
*c Etheostoma asprigene		X	X		X
*c E. chlorosomum		X	X		
*c E. gracile		X	X		
*c E. histrio	X			X	X
*c Percina caprodes	X	X	X		
*c P. sciera	X	X	X		
*c P. shumardi			X	X	X
Family Sciaenidae					
*c Aplodinotus grunniens		X	X		

Source: Adapted from monitoring zone 1 from Academy of Natural Sciences Report 9–96, 1995 Sabine River Studies for the Texas Eastman Company.

*Species found in the habitat described in the section "Structure of Aquatic Ecosystems."

[a] c, Carnivore; dh, detritivore-herbivore; do, detritivore-omnivore (devours organisms and all types of detritus); ho, herbivore-omnivore; o, omnivore.

[b] V, vegetation; M, mud; S, sand; G, gravel; R, rock.

N. minima, N. paratunkae, N. pelliculosa, and *N. pupula* var. *rectangularis. N. rhyn-chocephala* var. *germainii* was in among the plant debris, as were *N. salinicola, N. symmetrica,* and *N. texana.* Several species of *Pinnularia* were found associated with the plants. They were *Pinnularia biceps, P. intermedia,* and *P. major.* Growing in and among the plants in quiet water were *Bacillaria paradoxa, Denticula elegans,* and *Nitzschia amphibia.* Several species of *Nitzschia* were found in and among the vegetation (Table 1.6). These Nitzchias may on occasion grow attached to the plants, but typically are among the debris and roots of the plants. They are *Nitzschia acicularis, N. biacrula, N. circumsuta, N. clausii, N. frustulum, N. frustulum* var. *perminuta, N. mediastalsis, N. obtusa, N. obtusa* var. *scalpelliformis,* and *N. umblicata* (Table 1.6). On the debris was also found *Surirella salina.*

One of the monocotyledons present was *Fimbristylis vahlii.* These plants and their debris form excellent habitats for several species of animals. Among these were the protozoan, *Euglena* sp., an autotroph. Attached to the plants were an unidentified species of the Spongillidae and the detritivore-omnivore, *Ferrissia rivularis.* In and among the plants was the decapod, *Palaemonetes kadiakensis,* a herbivore-omnivore.

A few mayflies were found in this habitat. They were an unidentified species of *Baetis,* which is a herbivore-detritivore, perhaps an omnivore; *Isonychia sicca,* an omnivore; and an unidentified species of *Tricorythodes,* also an omnivore. A few Hemiptera were found: the carnivorous *Belostoma lutarium, Ranatra* sp., and *Rhagovelia* sp. Also present was the carnivorous *Corydalus cornutus.* A few beetles were found in this habitat: *Dineutes,* perhaps more than one species, and *Gyrinus* sp. Both of these are probably carnivores. A few caddisflies were found in this habitat. They were unidentified species of the genus *Hydroptila,* which are herbivores; and *Nectopsyche candida,* an omnivore. Several species belong to the genus *Cyrnellus,* which are omnivores. Several chironomids were in and among the debris of the vegetation. They were the omnivorous *Ablabesmyia* sp.; *A. rhamphe,* also an omnivore; *Glyptotendipes* sp., a herbivore; *Polypedilum,* an omnivore; and *Thienemannimyia* sp., omnivore. All of these were found in the detritus of the plants.

Several fish were found in and among the plants. They were the carnivorous *Lepisosteus oculatus* and *L. osseus.* Other fish that might be found in and among the vegetation were the omnivorous *Dorosoma cepedianum; Hybognathus nuchalis,* an onivore; and *Notropis atrocaudalis,* a carnivore. The omnivorous *Gambusia affinis* was also found grazing in and among the plants. Other fish found associated with the plants were the carnivorous *Lepomis humilis;* the omnivorous *L. macrochirus; L. megalotis,* a carnivore; and the carnivore *Pomoxis annularis.* Three species of percids were also found. They were the carnivorous *Etheostoma histrio, Percina caprodes, P. sciera,* and *P. shumardi.*

Lentic Habitats.   The lentic habitats were mainly the pools that had mud and sand as their substrate. Sand was more prevalent than mud in many cases. In the lentic habitats were *Amaranthus tamariscina, Mollugo verticillata, Ludwigia decurrens, Polygonum persicaria, Lindernia dubia,* and the composite *Eclipta alba.* Also floating on the surface of the water in the pool was an unidentified species of *Lemna.* In the edges of the pools or in the margins of the stream was *Fimbristylis vahlii.*

Several algae were found in these areas where the current was very slow or not present, and they were often attached to the mud and debris. They occurred not only in pools but also along the margins of the stream.

Several blue-green algae were quite common in this habitat. They were *Agmenellum quadruplicatum, Anabaena oscillariodes, Schizothrix arenaria, S. calcicola,* and *S. friesii.* The colonial green algae, *Scenedesmus ecornis* and *S. quadricauda,* were in this habitat. A few desmids were found in the pools: *Closterium acerosum, C. venus,* and *Spirogyra* sp.

These pools and shallow-water areas along the margins of the stream were favorite habitats for several species of diatoms: *Cyclotella aliquantula, C. atomus, C. comta, C. meneghiniana,* and *C. pseudostelligera.* Attached to debris and sometimes to plants living in this area were *Melosira distans* var. *alpigena.* Associated with the sandy mud were *Stephanodiscus astraea* var. *minutula, S. hantzschii, S. invisitatus,* and *S. minutus.* Living on the sandy mud was *Biddulphia laevis.* Attached to debris and plants living in these lentic areas was *Synedra ulna.* On the surface of the detritus in these lentic habitats were also *Eunotia exigua, E. major,* and *E. monodon.* Attached to the debris and stems of plants growing in these lentic habitats were *Achnanthes exigua, A. minutissima,* and *A. submontana.* Other diatoms in and among the debris were *Cymbella minuta, C. minuta* var. *silesiaca, C. naviculiformis,* and *Gomphonema clevei.* Other Gomphonemas in these habitats were *G. grunowii* and *G. parvulum,* which were attached to the debris in and among the substrate. Many species belonging to the order Naviculales and family Naviculaceae were found in these lentic habitats. On the surface of the sandy mud were *Anomoeoneis folis* and *Caloneis bacillum.* Also present on the surface of the substrate, muddy sand, were *Frustulia rhomboides* var. *crassinervia* and *F. weinholdii.* On the surface of the sandy mud were *Gyrosigma nodiferum, G. obscurum,* and *G. spencerii.* The genus *Navicula* was represented by 19 taxons in these lentic habitats (Table 1.6). Usually, they were either on the surface of the substrate or attached to the debris or plants living in these habitats. A single taxon belonging to the genus *Neidium* was found on the sandy mud. The genus *Pinnularia* was represented by four taxons living on the fine sand containing very little mud. The family Nitzschiaceae was well represented in this habitat. Moving over the substrate was *Bacillaria paradoxa.* In and among the debris were *Nitzschia amphibia, N. clausii, N. confinis, N. filiformis, N. frequens, N. frustulum* var. *perminuta,* and *N. frustulum* var. *subsalina.* Other Nitzschias associated with the debris were *N. hungarica, N. intermedia,* and *N. granulata.* On muddy substrates that were organically enriched were *N. palea, N. subtilis,* and *N. tarda.* The diatom *Surirella salina* was found in substrates consisting of mud, sand, and debris.

As stated earlier, a few plants were found in these lentic habitats: *Mollugo verticillata, Amaranthus tamariscina, Eclipta alba, Lemna* sp., and *Fimbristylis vahlii.* Several protozoans were found in these lentic habitats usually associated with the debris or plants. They were *Euglena* sp., an autotroph, and *Trachelomonas hispida,* also an autotroph. Attached to the debris was a detritivore-omnivore, *Ferrissia rivularis.* In the sand containing small amounts of mud were several unionids: *Amblema plicata, Arcidens confragosus, Megalonaias nervosa, Quadrula quadrula,*

Tritogonia verrucosa, Truncilla donaciformis, T. truncata, and *Utterbackia imbecillis.* All of these species are detritivore-herbivores. Also present was *Corbicula fluminea,* which is a filter feeder and therefore a detritivore-omnivore, as is *Sphaerium striatinum.*

Crawling over the debris was the tubificid worm, *Branchiura sowerbyi,* a detritivore-omnivore. Two crayfish, *Procambarus dupratzi* and *Orconectes palmeri* var. *longimanus,* were found in these lentic habitats in among the plants and debris.

Insects were also fairly common in these lentic habitats. There were a few odonates, the carnivorous *Argia* sp.; the carnivorous *Nasiaeschna pentacantha,* and the carnivorous *Macromia* sp. A few mayflies were present: the omnivorous *Hexagenia* sp. and *Tricorythodes* sp., also an omnivore. A few bugs were present: the carnivorous *Belostoma lutarium* and *Ranatra* sp. Present in and among the detritus was *Corydalus cornutus,* a carnivore, which belongs to the order Megaloptera. A very few beetles were found in this habitat. They were *Deronectes* sp., a carnivore and often associated with the plants and their debris. Another species found in this habitat was *Dineutes,* a surface diver that is carnivorous. Elmid beetles were also found in these lentic habitats. They were species belonging to the genus *Stenelmis,* which are omnivores. Several chironomids were in and among the mud and debris. They were *Ablabesmyia* sp. and *A. rhamphe,* both omnivores, and an undetermined species of the genus *Glyptotendipes,* which is a herbivore. Another chironomid belongs to the genus *Thienemannimyia,* which is an omnivore.

Several fish were found in the pools. They were the carnivores *Lepisosteus oculatus* and *L. 'osseus.* Also present was the omnivorous *Dorosoma cepedianum.* Also in the pools were the carnivores *Cyprinella lutrensis, C. venusta, Lythurus fumeus,* and *Macrhybopsis aestivalis* and the omnivore *Carpiodes carpio.* Various cyprinids were also found in the pools or grazing on the algae and debris along the margins of the stream where lentic conditions existed. They were *Hybognathus nuchalis* and the carnivores *Notropis amnis, N. atrocaudalis, N. buchanani,* and *N. sabinae.* Another species in the pools was *Pimephales vigilax,* an omnivore. Several catfish were found in these lentic habitats. They were *Ictalurus furcatus,* a carnivore; *I. punctatus,* an omnivore; *Notorus nocturnus,* a carnivore; and *Pylodictus olivaris,* a carnivore. A few cyprinids were found in these lentic habitats. They were *Gambusia affinis,* an omnivore, and *Labidesthes sicculus,* a carnivore. Belonging to the order Perciformes were *Lepomis humilis,* a carnivore; *L. macrochirus,* an omnivore; *L. megalotis,* a carnivore; *Micropterus punctulatus,* an omnivore; and *Pomoxis annularis,* a carnivore. Other fish were the Percidae, *Ammocrypta vivax,* a carnivore; *Etheostoma asprigene,* a carnivore; *E. chlorosomum,* a carnivore; *E. gracile,* a carnivore; *Percina caprodes,* a carnivore; *P. sciera,* a carnivore; *P. shumardi,* a carnivore; and *Aplodinotus grunniens,* a carnivore.

Lotic Habitats. The substrate of the lotic environment is mainly rocks, pebbles, and gravel. The current varies greatly depending on how the rocks are placed. Downstream from the riffle is the slack water. The bed of the slack water is mainly gravel and sand. Growing attached to the rocks were *Microcoleus vaginatus* and *Schizothrix calcicola.* Also attached to the rocks was the green alga, *Cladophora glomerata.*

Several diatoms were found among the algal mats and also in the gravel and sand between the rocks where the current was somewhat reduced. These diatoms were *Cyclotella meneghiniana, Melosira ambigua,* and *M. varians.* Other diatoms found attached to the gravel in the slack waters were *Synedra rumpens* var. *familiaris, Eunotia exigua, Achnanthes exigua,* and *A. minutissima.* Atached to the rocks in the riffle were specimens of *Cocconeis fluviatilis* and *Cymbella minuta.* Among the gravel is *Amphora submontana.* Also attached to the gravel in the riffles, but in areas where the current was moderate, were specimens of *Gomphonema clevei, G. grunowii,* and *G. intricatum.* In the slack waters attached to the pebbles was *Navicula contenta* var. *biceps. Navicula pelliculosa* was also in this habitat in areas where the current was less strong. In the slack water were *N. pupula, N. pupula* var. *mutata,* and *N. pupula* var. *rectangularis.* In very much reduced current, in between the gravel grains, were *Navicula tripunctata* var. *schizonemoides* and *Pinnularia biceps.* Several diatoms belonging to the family Nitzschiaceae were found in and among the gravel in the slack water. They were *Bacillaria paradoxa, Nitzschia apiculata, N. accomodata, N. bacata,* and *N. constricta* var. *subconstricta.* Also present were *N. gracilis, N. keutzingiana,* and *N. tarda.* Attached to the rocks where the current was moderately fast was the red alga, *Audouinella violacea.*

In and among the sand and gravel in the slack water were several protozoans. The autotroph, *Euglena* sp., was present. In and among the gravel in the slack water were several mussels: *Amblema plicata, Fusconaia lananensis, Lampsilis satura,* and *L. teres.* Other unionids found in these lotic habitats were *Quadrula pustulosa* var. *mortoni, Q. quadrula, Strophitus undulatus, Tritogonia verrucosa, Truncilla donaciformis,* and *T. truncata.* All of these are omnivore-herbivores; that is, they eat bacteria, protozoans, and various unicellular algae or colonial algae, which they engulf by filtration. The other unionid sometimes forming beds of single species was the detritivore-omnivore, *Corbicula fluminea.* A few crayfish were found in and among the rocks. They were *Procambarus dupratzi* and *Orconectes palmeri* var. *longimanus.*

Several insects were found on the rocks, often on the undersides of rocks or in areas where the current was somewhat reduced. They were the odonate, *Argia* sp., a carnivore, and the carnivores *Nasiaeschna pentacantha, Neocordulia molesta,* and *Macromia* sp. Also present was *Baetis* sp., detritivore-omnivore; *Stenacron interpunctatum,* an omnivore; and *S. exiguum,* also an omnivore, as is *Isonychia sicca.* A few stoneflies were found in this habitat. In the gravel, particularly associated with algae living in the gravel, was *Neoperla* prob. *clymene.* The coleopterans *Deronectes* spp. (carnivores) were found here. In gravel and rocks was *Rhagovelia* sp. Several Hydropsychidae were attached to the rocks. They were *Cheumatopsyche* sp., *Hydropsyche simulans, H. orris, Macrostemum carolina,* and *Potamyia flava,* all omnivores. A few chironomids were found in and among the gravel. They were an unidentified species of *Ablabesmyia,* an omnivore; *A. rhamphe,* an omnivore; *Rheotanytarsus* sp., an omnivore; and *Thienemannimyia* sp., an omnivore. Blackflies (*Simuliums* spp.) were also found here.

Attached to the rocks were the bryozoans *Plumatella emarginata* and *P. repens.* These are filter feeders and probably omnivores. The riffles and slack waters were favorite habitats for several species of fish. They were a carnivore, *Lepisosteus osseus; Dorosoma cepedianum,* an omnivore; *Cyprinella lutrensis,* a carnivore; *Lythurus fumeus,* a carnivore; *Macrhybopsis aestivalis,* a carnivore, and *Carpiodes carpio,* an omnivore. Also present were *Notropis buchanani,* a carnivore; *Ictalurus furcatus,* a carnivore; *I. punctatus,* an omnivore; and *Notorus nocturnus,* a carnivore; and *Fundulus notatus,* an omnivore. Particularly in the slack waters were *Labidesthes sicculus,* a carnivore; *Lepomis megalotis,* a carnivore; and *Micropterus punctulatus,* an omnivore. Three other species, *Ammocrypta vivax, Etheostoma histrio,* and *Percina shumardi,* all carnivores, were found among the rocks.

Lower Sabine.   In the summers of 1952 and 1970, studies of the lower Sabine resulted in the following association of aquatic organisms.

Vegetative Habitats.   In the vegetative habitats that existed along the edges of the stream, there were beds of various species of *Scirpus* and *Typha,* which were rooted in the shallow water, and *Phragmites,* which was in the very shallow water. Living partially submerged and partially emerged were various species of *Polygonum* and *Sagittaria.* Associated with this vegetation were *Microcoleus tenerrimus,* a blue-green alga and *Nostoc* sp., *Phormidium retzii* and *Phormidium subuliforme.* Also present were the green algae, *Rhizoclonium hieroglyphicum* and *Rhizoclonium* sp. In among the debris were *Chaetophora elegans,* the desmid *Closterium moniliferum,* and unidentified species of *Spirogyra* (Table 1.7).

Also in and among the debris of the vegetation and living attached to the stems of some of the emergent plants were *Melosira distans* var. *alpigena,* and in the debris were the diatoms *Hydrosera triquetra* and *Terpsinoe musica.* Also found were *Synedra goulardi* and *S. ulna. Synedra* sp. and *Eunotia praerupta* var. *bidens* were attached to some of the plant debris. Attached to some of the plant stems in the water was an unidentified species of the genus *Achnanthes.* Also present in this habitat was an unidentified species of the genus *Cocconeis. Caloneis bacillum* and *Gomphonema parvulum* were attached to the plant debris found in shallow waters, as was *Diploneis ovalis.* In and among the debris were three taxons belonging to the genus *Frustulia* and *Gyrosigma scalproides.* Also in this habitat were *Navicula confervacea* and *N. crucicula.* Other Naviculas associated with the vegetation were *Navicula mobiliensis* and *N. mutica.* Among the plant debris were found *Pinnularia acrosphaeria, Stauroneis crucicula, Nitzschia amphibia, N. clausii,* and *N. filiformis.* Other species of Nitzschias found in this habitat were *N. keutzingiana, N. obtusa,* and *N. sigma.* Also among the plant detritus was *N. tryblionella* var. *victoriae.* Also present in this habitat were *Surirella striatula* and *S. tenera* var. *nervosa.*

Two species of the order Peridinales, *Glenodinium cinctum* and *G. oculatum,* were also moving around among the plant detritus. It was surprising to find the red alga, *Compsopogon coeruleus,* in this habitat.

Plant debris and the base of plant stems were a favorite habitat for several species of protozoans. They were the flagellates *Anthophysis steinii, A. vegetans, Monas sociabilis,* and an unidentified species of *Monas.* Other flagellates found in this habitat were *Cryptomonas erosa, C. marssonii,* and *Cyathomonas truncata. Cyathomonas* and *Cryptomonas* are primary producers. Other primary producers in this habitat

(*text continues on page 54*)

TABLE 1.7. *Species List: Lower Sabine River*

Taxon[a]	V	M	S	G	R	Site[c]	Source[d]
		Lentic		Lotic			
		M	S	G	R		
SUPERKINGDOM PROKARYOTAE							
KINGDOM MONERA							
Division Cyanophycota							
Class Cyanophyceae							
Order Nostocales							
Family Nostocaceae							
* *Anabaena tortulosa*		X		X		O	a
Cylindrospermum lichenforme		X	X			O	b
Nostoc sp.	X	X				O	b
Family Oscillatoriaceae							
Lyngbya aeruginea acerulea		X		X		O	a
L. lagerheimii		X		X		O	a
* *L. lutea*		X		X		O	a
* *L. sordida*		X	X			O	a
Oscillatoria princeps		X		X	X	O	b
O. retzii		X	X	X	X	O	b
* *Microcoleus tenerrimus*	X		X	X	X	O	a
M. vaginatus		X		X	X	O	b
* *Phormidium retzii*	X	X		X	X	O	a
* *P. subuliforme*	X	X				O	a
P. valderianum				X	X	O	a
Schizothrix calcicola			X	X		O	b
S. mexicana				X	X	O	b
S. tenerrima				X	X	O	b
SUPERKINGDOM EUKARYOTAE							
KINGDOM PLANTAE							
Subkingdom Thallobionta							
Division Chlorophycota							
Class Chlorophyceae							
Order Ulothrichales							
Family Ulothrichaceae							
* *Ulothrix implexa*			X	X		O	a
Order Chaetophorales							
Family Chaetophoraceae							
* *Chaetophora elegans*	X					O	a
Order Oedogoniales							
Family Oedogoniaceae							
* *Oedogonium* spp.				X	X	O	a,b
Order Siphonocladales							
Family Cladophoraceae							
Pithophora oedogonia		X				O	b
* *Rhizoclonium hieroglyphicum*	X			X		O	a,b
* *Rhizoclonium* sp.	X					O	a
Order Zygnematales							

(*continued*)

TABLE 1.7.   (*Continued*)

Taxon[a]	V	Lentic M	Lentic S	Lotic G	Lotic R	Site[c]	Source[d]
Family Desmidiaceae							
* *Closterium moniliferum*	X	X				O	a
Family Zygnemataceae							
* *Spirogyra* spp.	X			X	X	O	a,b
Division Chromophycota							
Class Bacillariophyceae							
Order Eupodiscales							
Family Coscinodiscaceae							
Cyclotella aliquantula				X		O	b
C. atomus	X			X		O	b
C. florida	X	X	X	X		O	b
C. meneghiniana	X	X	X	X	X	O	a,b
C. pseudostelligera	X			X	X	O	b
* *Melosira distans* v. *alpigena*	X		X	X		O	b
M. granulata v. *angustissima*	X	X	X	X	X	O	b
M. italica v. *valida*	X	X		X	X	O	b
M. varians	X	X	X	X	X	O	a,b
Order Biddulphiales							
Family Biddulphiaceae							
* *Hydrosera triquetra*	X					O	b
* *Terpsinoe musica*	X			X		O	a,b
Order Fragilariales							
Family Fragilariaceae							
Asterionella formosa				X	X	O	b
Dimerogramma lanceolatum						O	b
Fragilaria construens v. *subsalina*				X	X	O	b
F. pinnata	X	X	X			O	b
F. virescens v. *clavata*						O	b
Opephora marina	X	X	X			O	b
O. martyii			X	X	X	O	b
Synedra affinis	X					O	b
S. affinis v. *obtusa*	X	X				O	b
* *S. goulardi*	X			X		O	a
S. pulchella				X		O	b
S. rumpens	X			X		O	b
S. tenera				X		O	b
* *S. ulna*	X	X	X	X		O	a,b
S. ulna v. *danica*	X			X		O	a
S. ulna v. *oxyrhynchus*	X			X		O	b
S. vaucheriae						O	a,b
* *Synedra* sp.	X		X			O	a
Order Eunotiales							
Family Eunotiaceae							
* *Eunotia formica*				X		O	a
E. lunaris				X		O	a,b

TABLE 1.7.   (*Continued*)

Taxon[a]	V	Lentic M	Lentic S	Lotic G	Lotic R	Site[c]	Source[d]
Family Eunotiaceae (*cont.*)							
E. monodon v. major	X			X		O	b
E. pectinalis v. minor	X			X		O	b
* E. praerupta v. bidens	X			X		O	b
E. tenella				X	X	O	b
Order Achnanthales							
Family Achnanthaceae							
Achnanthes biporoma	X			X	X	O	b
A. exigua v. heterovalvata	X	.	X	X		O	b
A. hauckiana				X	X	O	b
A. hauckiana v. rostrata				X		O	b
A. lanceolata	X	X		X		O	a,b
A. longiceps						O	b
A. minutissima	X			X	X	O	a,b
* A. temperei		X				O	a
A. tenera						O	b
* Achnanthes sp.	X			X		O	a
Cocconeis fluviatilis					X	O	b
C. placentula	X			X	X	O	b
C. placentula v. lineata	X			X	X	O	b
* Cocconeis sp.	X			X	X	O	a
Order Naviculales							
Family Cymbellaceae							
Amphora cymbelloides	X			X	X	O	a
A. ovalis	X			X		O	b
* Amphora sp.				X		O	b
* Cymbella tumida				X		O	b
C. ventricosa		X	X	X		O	a,b
Family Gomphonemaceae							
Gomphonema brasiliense	X	X	X	X		O	b
G. gracile						O	b
* G. gracile v. naviculoides		X	X			O	a
G. intricatum v. pumila			X	X		O	b
G. longiceps v. subclavata						O	b
* G. parvulum	X	X	X	X		O	a,b
G. sphaerophorum						O	b
Gomphonema sp.						O	a
Family Naviculaceae							
Amphipleura pellucida			X	X		O	a
Anonmeoneis exilis		X				O	a
* Caloneis bacillum	X			X		O	a,b
Diploneis oblongella	X	X				O	b
* D. ovalis	X					O	b
D. pseudovalis	X					O	b
D. puella						O	a,b
D. smithii						O	b

(*continued*)

TABLE 1.7. (*Continued*)

Taxon[a]	Lentic			Lotic		Site[c]	Source[d]
	V	M	S	G	R		
Family Naviculaceae (*cont.*)							
* *Frustulia rhomboides*	X	X		X		O	a,b
* *F. rhomboides* v. *crassinervia*	X	X		X		O	a,b
F. vulgaris						O	b
* *F. weinholdii*	X	X				O	a
Gyrosigma acuminatum						O	b
G. attentuatum						O	b
* *G. scalproides*	X	X		X		O	a
G. spencerii v. *nodifera*					X	O	a
Navicula capitata v. *hungarica*					X	O	b
N. cincta v. *leptocephala*						O	a
* *N. confervacea*	X	X	X	X		O	a
N. contenta f. *biceps*						O	b
* *N. crucicula*	X	X				O	a
N. cryptocephala					X	O	b
N. deserta						O	b
N. expansa						O	a
N. germainii		X				O	b
N. graciloides		X	X			O	b
N. ingenua						O	b
N. integra						O	b
N. minima	X			X		O	b
* *N. mobiliensis*	X	X				O	a
* *N. mutica*	X		X	X		O	a
N. mutica v. *cohnii*				X		O	b
N. paratunkae		X	X	X		O	b
N. pupula	X			X		O	b
N. pupula v. *capitata*	X			X		O	b
N. pupula v. *minutula*	X			X		O	b
N. radiosa	X		X	X		O	a
N. rectangularis						O	b
* *N. schroeteri*			X	X		O	a
N. schroeteri v. *excambia*				X		O	b
N. secreta v. *apiculata*				X		O	b
N. seminulum				X		O	b
* *N. subatomoides*		X	X			O	a
N. symmetrica	X			X		O	a,b
N. tenera				X		O	b
N. tripunctata v. *schizonemoides*				X		O	b
* *Navicula* spp.		X	X			O	a
* *Pinnularia acrosphaeria*	X	X				O	a
P. braunii v. *amphicephala*		X	X			O	b
P. interrupta	X					O	b
* *Stauroneis crucicula*	X	X	X			O	a
Order Epithemiales							

TABLE 1.7. (*Continued*)

Taxon[a]	Substrate[b]					Site[c]	Source[d]
	Lentic			Lotic			
	V	M	S	G	R		
Family Epithemiaceae							
* *Rhopalodia gibba*				X		O	a
* *R. gibberula*				X		O	b
Order Bacillariales							
Family Nitzschiaceae							
Bacillaria paradoxa				X	X	O	a,b
Gomphonitzschia agma						O	b
* *Nitzschia amphibia*	X					O	a
N. bacata	X	X	X			O	b
* *N. brevissima*				X		O	a,b
* *N. clausii*	X		X	X		O	a,b
* *N. fasciculata*				X		O	a,b
* *N. filiformis*	X	X	X	X		O	a,b
* *N. fonticola*				X		O	a
N. frustulum	X		X	X		O	b
N. frustulum v. *indica*				X		O	b
N. frustulum v. *perminuta*				X	X	O	b
N. ignorata						O	b
* *N. kutzingiana*	X	X		X		O	a,b
* *N. lorenziana* v. *subtilis*				X		O	a,b
N. microcephala						O	b
* *N. obtusa*	X			X		O	b
N. palea	X	X	X	X	X	O	a,b
* *N. sigma*	X	X				O	b
N. tarda		X	X			O	b
N. tropica		X	X			O	b
N. tryblionella v. *debilis*				X		O	b
* *N. tryblionella* v. *levidensis*				X		O	b
* *N. tryblionella* v. *victoriae*	X	X		X		O	a
Nitzschia spp.						O	a
Order Surirellales							
Family Surirellaceae							
Surirella brightwellii		X	X			O	a
S. minuta		X	X			O	a
S. ovata						O	b
S. ovata v. *salina*						O	a
* *S. striatula*	X	X				O	a
* *S. tenera* v. *nervosa*	X					O	a
unident. sp.						O	b
Class Dinophyceae							
Order Gymnodiniales							
Family Gymnodiniaceae							
Gymnodinium palustre	X					O	b
Gymnodinium sp.	X					O	b
Gyrodinium hyalinum						O	b
Order Peridinales							

(*continued*)

TABLE 1.7.   (*Continued*)

Taxon[a]	V	M	S	G	R	Site[c]	Source[d]
		Lentic		Lotic			
		Substrate[b]					
Family Cystodiniidae							
* *Glenodinium cinctum*	X					O	a
* *G. oculatum*	X					O	a
Division Rhodophycota							
Class Rhodophyceae							
Order Compsopogonales							
Family Compsopogonaceae							
* *Compsopogon coeruleus*	X				X	O	a,b
Order Acrochaetinales							
Family Audouinellaceae							
Audouinella violacea					X	O	a
Order Nemalionales							
Family Batrachospermaceae							
Batrachospermum sp.					X	O	b
SUBKINGDOM EMBRYOBIONTA							
Division Magnoliophyta							
Class Dicotyledoneae							
Order Nymphaeales							
Family Cambombaceae							
Cambomba caroliniana	X					O	t
Family Ceratophyllaceae							
* *Ceratophyllum demersum*		X				O	t
Order Polygonales							
Family Polygonaceae							
* *Polygonum* spp.	X	X				O	t
Order Apiales							
Family Apiaceae							
(= *Umbelliferaceae*)							
* *Hydrocotyl* sp.		X				O	t
Class Monocotyledoneae							
Order Alismales							
Family Alismaceae							
* *Sagittaria* spp.	X	X				O	t
Order Hydrocharitales							
Family Hydrocharitaceae							
* *Vallisneria americana*		X	X			O	t
Order Najadales							
Family Potamogetonaceae							
* *Potamogeton* sp.		X	X			O	t
Order Arales							
Family Lemnaceae							
Lema spp.						O	t
Order Cyperales							
Family Cyperaceae							
* *Scirpus* spp.	X					O	t
unident. spp.						O	a,t

TABLE 1.7. (*Continued*)

Taxon[a]	V	Lentic M	S	Lotic G	R	Site[c]	Source[d]
Family Poaceae							
* *Phragmites* sp.	X					O	a
Order Typhales							
Family Typhaceae							
* *Typha* spp.	X					O	t
Order Iliales							
Family Pontederiaceae							
Eichhornia crassipes						O	a,t
KINGDOM ANIMALIA							
SUBKINGDOM PROTOZOA							
Class Mastigophora							
Order Chrysomonadida							
Family Ochromonadidae							
* *Anthophysis steinii*	X					O	a
* *A. vegetans*	X					O	a,b
Monas elongata	X					O	b
M. guttula	X					O	b
* *M. sociabilis*	X					O	b
* *Monas* sp.	X					O	a
Order Cryptomonadida							
Family Cryptomonadidae							
Chilomonas paramecium		X	X			O	a,b
*o *Cryptomonas erosa*	X					O	a,b
*o *C. marssonii*	X	X				O	a
*o *Cyathomonas truncata*	X	X				O	a
Order Phytomonadida							
Family Carteriidae							
*o *Carteria cordiformis*	X	X				O	a
C. globosa	X	X				O	b
Order Volvocales							
Family Chlamydomonadaceae							
Chlamydomonas globosa	X					O	b
C. inversa	X					O	a
C. monadina	X					O	b
C. variabilis	X					O	a
*o *Chlamydomonas* sp.	X					O	a
Diplostauron sp.	X					O	a
*o *Scourfieldia complanata*	X					O	a
Family Volvocaceae							
Eudorina elegans	X					O	b
*o *E. illinoiensis*	X					O	a
*o *Gonium pectorale*	X					O	b
*o *Pandorina morum*	X					O	a,b
Platydorina caudata						O	b
Pleodorina illinoiensis						O	b
Spondylomorum sp.						O	b
Order Euglenoidida							

(*continued*)

TABLE 1.7.   (*Continued*)

		Lentic		Lotic			
Taxon[a]	V	M	S	G	R	Site[c]	Source[d]
Family Anisonemidae							
*o *Anisonema acinus*	X					O	a
A. emarginatum	X					O	b
*o *Anisonema* sp.	X					O	a
*o *Entosiphon ovatum*	X					O	a
Heteronema acus	X	X				O	b
Notosolenus apocamptus	X	X				O	b
*o *N. lens*	X	X				O	a
*o *Peranema deflexum*	X	X				O	a
*o *P. trichophorum*	X	X				O	a,b
Family Astasiidae							
Petalomonas quadrilineata	X					O	a
Urceolus cyclostomus	X	X				O	a
Family Euglenidae							
*o *Ascoglena* sp.	X					O	a
Euglena agilis	X					O	a
*o *E. anabaena*	X					O	a
E. caudata	X					O	a
*o *E. fusca*	X					O	a
*o *E. haematodes*	X					O	a
*o *E. intermedia*	X					O	a
E. oxyuris	X					O	b
E. pisciformis	X					O	a
E. terricola	X					O	b
* *E. viridis*	X					O	a
*o *Lepocinclis texta*		X				O	a
Phacus accuminata		X				O	b
P. alata		X				O	a
P. anacoelus		X				O	a
*o *P. curvicauda*		X				O	a
P. longicauda		X				O	b
*o *P. orbicularis*		X				O	a
P. pleuronectes		X				O	b
P. torta		X				O	b
Phacus sp.		X				O	a
Trachelomonas hispida	X	X				O	a,b
T. horrida	X	X				O	a
T. mirabilis	X	X				O	b
*o *T. schauinslandii*	X	X				O	a
*o *T. verrucosa*	X	X				O	a,b
*o *T. volgensis*	X	X				O	a
*o *T. volvocina*	X	X				O	a
Order Protomonadida							
Family Bodonidae							
d *Bodo allexeieffii*	X	X				O	a
d *B. caudatus*	X	X				O	b
d *B. obovatus*	X	X				O	a

TABLE 1.7. (*Continued*)

Taxon[a]	Substrate[b]					Site[c]	Source[d]
	V	Lentic		Lotic			
		M	S	G	R		
Family Bodonidae (*cont.*)							
d *B. variabilis*	X	X				O	b
Class Sarcodina							
Order Amoebida							
Family Amoebidae							
Amoeba gorgonia	X	X				O	b
A. guttula	X	X				O	b
A. radiosa	X	X				O	b
A. spumosa	X	X				O	a
A. striata	X	X				O	a
Vahlkampfia limax	X	X	X			O	a,b
Order Testacida							
Family Arcellidae							
*do *Arcella gibbosa*	X	X				O	a
A. vulgaris	X					O	b
*do *Arcella* sp.	X	X				O	a
Family Difflugiidae							
*do *Difflugia arcula*	X	X				O	a
* *D. globosa*	X	X				O	b
D. oblonga	X	X				O	a
* *Difflugia* sp.	X	X				O	a
Order Heliozoida							
Family Acanthocystidae							
Acanthocystis sp.	X					O	b
Family Actinophryidae							
Actinophrys sol		X				O	b
Actinosphaerium eichhorni		X				O	b
Family Lithocollidae							
Lithocolla globosa		X				O	b
Class Ciliata							
Order Gymnostomatida							
Family Amphileptidae							
do *Amphileptus claparedie*	X					O	b
do *Lionotus fasciola*		X				O	b
do *L. trichocystus*		X				O	b
*o *Lionotus* sp.		X				O	a
do *Loxophyllum meleagris*						O	b
Family Chlamydodontidae							
d *Chilodonella cucullulus*	X	X	X			O	b
d *C. fluviatilis*	X	X	X			O	b
d *C. uncinata*	X	X	X			O	b
Family Holophryidae							
Holophrya sp.						O	b
Lacrymaria olor						O	b
Placus ovum						O	a
Platyophrya sp.						O	a

(*continued*)

TABLE 1.7.   (*Continued*)

Taxon[a]		V	M	S	G	R	Site[c]	Source[d]
			Lentic		Lotic			
							Substrate[b]	
Family Tracheliidae								
d	*Trachelius ovum*		X				O	b
	Order Trichostomatida							
	Family Colpodidae							
	Colpoda sp.						O	a
	Order Hymenostomatida							
	Family Frontoniidae							
*o	*Cinetochilum margaritaceum*		X				O	a,b
*o	*Cyrtolophosis musicola*		X					
	Frontonia elliptica		X				O	a
*o	*F. leucas*		X				O	b
	Family Parameciidae							
	Paramecium aurelia					X	O	a,b
*o	*P. multimicronucleatum*					X	O	a
	Family Pleuronematidae							
*o	*Cyclidium glaucoma*		X				O	a
*o	*C. hepatrichum*		X				O	a
	C. litomesum		X				O	b
	Order Heterotrichida							
	Family Stentoridae							
	Stentor igenus					X	O	b
	Order Oligotrichida							
	Family Halteriidae							
*o	*Halteria grandinella*	X			X	X	O	b
	Family Strobilidiidae							
*o	*Strobilidium concicum*	X				X	O	a
	Order Tintinnida							
	Family Tintinnidae							
o	*Tintinnidium semiciliatum*	X					O	a
	Order Hypotrichida							
	Family Aspidiscidae							
*o	*Aspidisca costata*		X				O	a,b
*o	*A. lynceus*		X				O	a
	Family Euplotidae							
*o	*Euplotes affinis*		X				O	a
*o	*E. carinatus*		X				O	a
*o	*E. novemcarinatus*		X				O	a
*o	*Euplotes* sp.		X				O	a
	Family Oxytrichidae							
*o	*Keronopsis spectabilis*		X				O	a
	Oxytricha setigera				X		O	b
	Stichotricha secunda				X		O	b
	Stylonychia mytilus				X		O	b
	S. pustulata				X		O	b
*d	*Uroleptus limnetis*	X			X		O	a
	Urosoma caudata				X		O	b
	Order Odontostomatida							

TABLE 1.7.   (*Continued*)

Taxon[a]	Substrate[b]					Site[c]	Source[d]
		Lentic		Lotic			
	V	M	S	G	R		
Family Epalxellidae							
Order Peritrichida							
Family Epistylidae							
Campanella umbellaria						O	b
Family Vaginicolidae							
do *Cothurnia annulata*						O	b
do *C. imberis*						O	b
do *C. ovata*						O	b
do *Cothurnia* sp.						O	b
do *Vaginicola annulata*						O	b
Family Vorticellidae							
do *Vorticella convallaria*	X			X		O	b
do *V. elongata*	X			X		O	b
do *V. microstoma*	X			X		O	b
do *V. similis*	X			X		O	b
*do *V. vestita*	X			X		O	a
Class Suctoria							
Order Suctoria							
Family Dendrosomatidae							
do *Trichophrya* sp.						O	b
Family Podophryidae							
do *Metacineta mystacina*					X	O	b
SUBKINGDOM PARAZOA							
Phylum Porifera							
Class Demospongiae							
Order Haplosclerida							
Family Spongillidae							
o *Spongilla fragilis?*	X	X		X		O	b
*o *Spongilla* cf. *lacustris*	X	X		X		O	a
do *Trochospongilla horrida*	X	X		X		O	a
SUBKINGDOM EUMETAZOA							
Phylum Cnideria							
Class Hydrozoa							
Order Hydroida							
Family Hydridae							
do *Hydra* sp.	X			X		O	s
Phylum Platyhelminthes							
Class Turbellaria							
Order Tricladida							
Family Planariidae							
do *Cura foremanii*						N	s
do *Dugesia* sp.	X	X		X		O	s
Phylum Nemertea							
unident. sp.						O	a
Class Enopla							
Order Hoplonemertea							

(*continued*)

TABLE 1.7. (*Continued*)

Taxon[a]	Substrate[b] V	Lentic M	Lentic S	Lotic G	Lotic R	Site[c]	Source[d]
Family Tetrastemmatidae							
*do *Prostoma rubrum*	X					N,O	s
Phylum Rotifera							
Class Digonota							
Order Bdelloidea							
Family Philodinidae							
Dissotrocha aculeata	X					O	a
Class Monogononta							
Order Ploima							
Family Brachionidae							
*o *Brachionus angularis*	X					O	a
*o *B. havanaensis*	X					O	a
*o *B. plicatilis*	X					O	a
B. quadridentatus	X					O	a
Kellicottia bostoniensis	X					O	a
*o *Keratella cochlearis*	X					O	a
K. gracilenta	X					O	a
*o *Platyais patulus*	X					O	a
Family Lecaniidae							
*o *Lecane luna*	X					O	a
Family Notommatidae							
Resticula nyssa	X					O	a
Family Synchaetidae							
*o *Polyarthra remata*	X					O	a
P. vulgaris	X					O	a
Order Floscialariaceae							
Family Filiniidae							
*o *Filinia terminalis*	X					O	a
*o *Pedalia jenkinae*	X					O	a
Phylum Nemata							
unident. spp.						O	s
Phylum Mollusca							
Class Gastropoda							
Order Mesogastropoda							
Family Hydrobiidae							
o *Amnicola limosa*	X			X	X	N,O	s
o *Amnicola* sp.	X			X	X	O	s
unident. spp.						O	b
Family Viviparidae							
o *Viviparus (Cochliopa)* sp.	X					O	s
Order Basommatophora							
Family Ancylidae							
o *Ferrissia* sp.					X	O	a,s
o *Hebetancylus excentricus*	X		X	X	X	O	b
o *Laevapex diaphanus*	X			X	X	N,O	s
unident. sp.						O	s

TABLE 1.7. (*Continued*)

Taxon[a]		Substrate[b]				Site[c]	Source[d]
	V	Lentic		Lotic			
		M	S	G	R		
Family Physidae							
o *Physa forsheyi*	X	X		X		O	b
o *P. virgata*	X	X		X		O	s
o *Physa* sp.	X	X		X		N,O	s
Family Planorbidae							
o *Helosoma trivolvis*	X	X				O	b
o *Menetus dilatatus*	X	X		X		N,O	s
o *M.? sampsoni*	X	X		X		O	b
Order Prosobranchiata							
Family Amnicolidae							
*o *Littoridina* sp.	X	X		X		O	a
Class Bivalvia							
Order Unionoida							
Family Unionidae							
o *Tritogonia* sp.		X				O	s
Order Veneroida							
Family Corbiculidae							
*o *Corbicula manilensis*	X	X	X			N,O	s
(= *fluminea?*)							
Family Dreissensiidae							
o *Cogeria leucophaeata*				X		O	b
*o *Mytilopsis leucophaeatus*				X		O	a
(= *Cogeria leucophaeata?*)							
Family Mactridae							
*o *Rangia cuneata*	X					O	a,b,s
Rangia sp.	X					O	a
Family Sphaeriidae							
*o *Eupera cubensis*	X			X		O	s
o *Musculum partumeium*		X	X			N,O	s
o *Pisidium* sp.		X	X			O	s
Phylum Annelida							
Class Polychaeta							
Order Phyllodocida							
Family Nereidae							
do *Laeonereis culveri*						O	a
Namalycastis sp.						O	b
Class Hirudinoidia							
Order Rhynchobdellida							
Family Glossiphoniidae							
c *Batracobdella phalera*		X				O	s
c *B. picta*		X				N	s
c *Helobdella elongata*	X					O	s
c *Placobdella hollensis*	X					N	s
c *P. monitifera*	X					O	s
c *P. multilineata*	X					O	s
c *Placobdella* sp.	X			X		O	s

(*continued*)

TABLE 1.7.   (*Continued*)

Taxon[a]	V	Lentic M	S	Lotic G	R	Site[c]	Source[d]
Family Piscicolidae							
unident. sp.							
Order Arhynchobdellida							
Family Erpobdellidae							
Dina? anoculata						O	b
Class Oligochaeta							
Order Haplotaxida							
Family Enchytraeidae							
unident. sp.						N	s
Family Lumbricidae							
do *Eclipidrilus* sp.						N,O	s
do *Eisenella tetraedra*						O	b
do *Lumbriculus variegatus*	X					O	s
do *Rhynchelmis* sp.						O	s
Family Megascolecidae							
unident. sp						O	b
Family Naididae							
do *Chaetogaster diastrophus*	X	X		X		O	s
do *Dero furcata*	X	X				N	s
(= *Aulophorus furcatus*)							
do *D. nivea*	X					O	s
do *D. obtusa*	X					O	s
*do *Dero* sp.	X					N,O	s
do *Dero (Aulophorus)* sp.	X					O	a
do *Haemonais waldvogeli*		X				O	s
do *Nais communis*	X					N,O	s
do *N. eliguis*	X	X				O	s
do *N. variabilis*	X					O	s
do *Ophidonais serpentina*						N	s
do *Paranais littoralis*						O	s
do *Pristina aequiseta*	X					N,O	s
do *P. breviseta*						N,O	s
do *P. idrensis*	X					O	s
do *P. longidentata*						O	s
do *P. longiseta*						O	s
do *P. longiseta leidyi*						O	s
do *P. longisoma*						N	s
P. osborni						N,O	s
do *P. plumaseta*	X					N,O	s
do *P. sima*						O	s
do *P. undentata*						O	s
do *Pristina* sp.						O	s
do *Slavina appendiculata*						N,O	s
do *Stylaria fossularis*		X				O	s
S. litoralis		X				O	s
*do *Stylaria* sp.		X				N	s
Order Branchiobdellida							

TABLE 1.7. (*Continued*)

Taxon[a]	Substrate[b]					Site[c]	Source[d]
		Lentic		Lotic			
	V	M	S	G	R		
Family Branchiobdellidae							
unident. sp.							
Family Tubificidae							
do *Aulodrilus pigueti*						N,O	s
do *A. pluriseta*	X	X				O	s
do *Aulodrilus* sp.						O	s
do *Branchiura sowerbyi*	X	X				N,O	s
do *Limnodrilus hoffmeisteri*	X	X				O	s
do *Limnodrilus* sp.	X	X				N,O	s
do *Peloscolex* (= *Quistadrilus*) *multisetosus*	X	X	X			N,O	s
do *Potamothrix vehodovskyi*						O	s
do *Rhyacodrilus coccineus*						N,O	s
do *Tubifex* sp.	X	X				N,O	s
unident. spp.						O	s
Phylum Arthropoda							
Class Arachnida							
Order Acariformes							
Family Athienamannidae							
unident. sp.						O	b
Family Pionidae							
c *Forelia* sp.	X		X	X		O	b
Family Tyrrelliidae							
unident. sp.						O	b
Family Unionicolidae							
Unionicola sp.	X					O	s
unident. sp.						O	s
Class Crustacea							
Subclass Cephalocarida (= Branchiopoda)							
Order Cladocera							
Family Chydoridae							
Eurycercus lamellatus	X		X	X		O	s
Family Daphnidae							
Daphnia sp.	X		X	X		O	s
Simocephalus sp.	X		X	X		O	s
Family Sididae							
Latona setifera	X		X	X		O	s
Sida crystallina	X		X	X		O	s
Subclass Copepoda							
Order Cyclopoida							
unident. spp.						O	s
Family Cyclopodae							
o *Cyclops* sp.						O	s
o *Macrocyclops* sp.						O	s
Subclass Cirripedia							

(*continued*)

TABLE 1.7. (*Continued*)

Taxon[a]	V	M	S	G	R	Site[c]	Source[d]
		Lentic		Lotic			
Order Thoracica							
Family Balanidae							
*o *Balanus amphitrite niveus*				X		O	a
o *Balanus* sp.				X		O	a
Subclass Malacostraca							
Order Mysidacea							
Family Mysidae							
o *Taphromysis louisianae*	X			X	X	N,O	b,s
Order Isopoda							
Family Aselidae							
Asellus intermedius	X			X		O	s
Lirceus louisianae	X	X		X		O	s
Family Parasellidae							
*do *Munna* sp.	X					O	a
Family Sphaeromidae							
o *Sphaeroma destructor*				X		O	b
Order Amphipoda							
Family Corophidae							
*o *Corophium lacustre*		X		X	X	O	b
Corophium cf. *lacustris*		X				O	a
Family Crangomyctidae							
o *Crangonyx* sp.				X	X	N,O	s
Family Gammaridae							
*o *Gammarus fasciatus*	X	X		X		O	a,s
o *G. mucronatus*	X			X		O	b
o *G. tigrinus*	X	X		X		O	b
o *Gammarus* spp.	X	X		X		N,O	s
Family Talitridae							
o *Hyalella azteca*	X	X	X	X	X	N,O	b,s
Order Decapoda							
Family Cambaridae							
o *Cambarellus schufeldti*	X	X	X		X	O	b
o *Cambarellus* sp.	X	X	X		X	O	a
Family Ocypodidae							
o *Uca pugnax*		X					
* *U. rapax*		X				O	b
Family Palaemonidae							
o *Macrobrachium ohione*	X	X	X			N,O	s,a
o *Palaemonetes kadiakensis*	X					O	b,s
*o *P. paludosus*	X		X	X	X	O	b
*o *P. pugio*	X		X	X	X	O	a
*o *P. vulgaris*	X		X	X	X	O	s
Family Penaeidae							
*o *Panaeus setiferous*	X	X		X	X	O	b

TABLE 1.7. (*Continued*)

Taxon[a]	V	M	S	G	R	Site[c]	Source[d]
		Lentic		Lotic			
Family Portunidae							
Callinectes sapidus		X				O	b
*c *C. sapidus acutidens*		X				O	a
Family Xanthidae							
*c *Rithropanopeus harrissi*	X	X		X	X	O	b
Class Insecta							
Order Collembola							
Family Isotomidae							
Isotomurus palustris	X	X	X			N	s
Order Odonata							
Suborder Zygoptera							
Family Calopterygidae							
c *Hetaerina* sp.				X	X	O	b,s
Family Coenagrionidae							
*c *Argia* sp.	X	X	X	X	X	N,O	s
* *Chromagrion* sp.	X	X	X	X	X	N	s
*c *Enallagma* nr. *signatum*	X	X	X			O	a
*c *Enallagma* sp.	X	X	X			O	b
c *Ischnura* nr. *posita*	X					O	a
c. *I.* nr. *verticalis*	X					O	a
Family Protoneuridae							
c *Neoneura* sp.				X		O	s
Suborder Anisoptera							
Family Aeshnidae							
*c *Nasiaeschna* sp.		X	X	X		N	s
Family Corduliidae							
c *Epicordulia* nr. *princeps*	X	X	X			O	b
c *Epicordulia* sp.	X	X	X			O	s
c *Somatochlora* sp.		X	X			N,O	s
Family Gomphidae							
c *Dromogomphus* sp.	X	X		X		N,O	s
c *Gomphus submedianus*	X	X				N	s
c *Gomphus* sp.	X	X				N,O	s
c *Octogomphus* sp.	X	X				O	s
c *Ophiogomphus* sp.	X			X	X	N	s
c *Progomphus* sp.	X			X	X	N,O	s
unident. sp						O	b
Family Libelludidae							
c *Libellula* sp.	X	X				N	s
c *Macrothemis* sp.	X	X				N	s
c unident. sp. nr. *erythemis*		X				O	b
c unident. sp. nr. *pachydiplax*		X				O	b
Family Macromiidae							
c *Didymops* sp.		X				N,O	s
c *Macromia* sp.		X				N,O	s
Order Ephemeroptera							

(*continued*)

TABLE 1.7.   (*Continued*)

Taxon[a]	V	M	S	G	R	Site[c]	Source[d]
		Lentic		Lotic			
	V	M	S	G	R	Site[c]	Source[d]
Suborder Schistonota							
Family Baetidae							
o *Baetis* sp.	X	X	X	X	X	N,O	b,s
o *Centroptilum* sp.	X	X	X	X	X	N,O	s
Family Ephemeridae							
o *Hexagenia bilineata*		X	X			O	s
o *H. limbata*		X	X			N,O	s
o *Hexagenia* sp.		X	X			N,O	a,s
Family Heptageniidae							
Cinygma (Heptagenia) sp.				X	X	O	s
o *Heptagenia diabasis*				X	X	N	s
o *Stenonema exiguum*				X	X	N	s
S. (= Stenacron) floridense				X	X	O	s
o *S. integrum*				X	X	N,O	s
o *Stenonema* spp.				X	X	O	a
Family Siphlonuridae							
o *Isonychia* sp.				X	X	N,O	a,s
Suborder Pannota							
Family Baetiscidae							
o unident. sp.	X					O	s
Family Caenidae							
o *Brachycercus* sp.		X	X			N,O	s
o *Caenis* sp.		X	X			N,O	s,b
Family Ephemerellidae							
o *Ephemerella temporalis*	X			X	X	N	s
Family Tricorythidae							
o *Tricorythodes* sp.	X	X	X			N,O	s
Order Plecoptera							
Family Perlidae							
Acroneuria sp.				X	X	N	s
c *Neoperla clymene*				X	X	N	s
c *Perlesta placida*	X	X	X	X	X	N,O	s
Order Hemiptera							
Family Belostomatidae							
c *Belostoma* nr. *lutarium*	X	X	X			O	b
c *Belostoma* sp.	X	X	X			O	b,s
Family Corixidae							
o *Sigara* sp.		X	X			N	s
Family Gerridae							
*c *Rheumatobates hungerfordi*	X			X	X		
c *Trepobates* sp.		X		X	X	N,O	a,s
Family Naucoridae							
c *Ambrysus?* sp.	X	X	X			O	b
c *Pelocoris femoratus*	X					O	b

TABLE 1.7. (*Continued*)

Taxon[a]		Substrate[b]					Site[c]	Source[d]
			Lentic		Lotic			
		V	M	S	G	R		
	Family Nepidae							
c	*Ranatra nigra*	X	X	X			O	b
	Family Notonectidae							
c	*Notonecta* sp.	X	X	X			O	a
	Order Megaloptera							
	Family Corydalidae							
c	*Corydalus cornutus*		X	X	X	X	N,O	s
	Family Sialidae							
c	*Sialis americana*	X	X	X	X	X	O	s
c	*S. velata*	X	X	X	X	X	O	s
c	*Sialis* sp.	X	X	X	X	X	N,O	s
	Order Coleoptera							
	Suborder Adephaga							
	Family Dytiscidae							
	Coelambus spp.	X					O	a
co	*Hydroporus* sp.	X	X	X			N	s
	Hydrovatus sp.	X	X	X			O	s
	Laccophilus sp.	X	X	X			O	b
	unident. sp. nr. *laccophilus*	X	X	X			O	b
	Family Gyrinidae							
c	*Gyretes* sp.	X					N	s
u	*Dineutus assimilis*	X					O	a
	D. nr. *discolor*	X					O	b
u	*Dineutus* sp.	X					N	s
o	*Gyrinus analis*	X					O	a
o	*G.* nr. *analis*	X					O	b
o	*G. lugens*	X					O	a
	Family Noteridae							
o	*Colpius* sp.	X	X	X			N	s
o	*Pronoterus* sp.	X					N	s
	unident. sp. nr. *suphisellus*	X	X	X			O	b
	Suborder Polyphaga							
	Family Brentidae							
	unident. sp.						N	s
	Family Curculionidae							
	Conotrachelus nenuphar	X	X	X			O	s
	Family Elmidae							
o	*Dubiraphia* sp.		X	X	X	X	N,O	s
o	*Narpus* sp.				X	X	N,O	s
o	*Stenelmis* nr. *quadrimaculata*		X	X	X	X	O	b
o	*Stenelmis* sp.		X	X	X	X	N,O	s
	unident. sp.						O	s
	Family Helodidae							
	unident. sp.						N	s

(*continued*)

TABLE 1.7. *(Continued)*

Taxon[a]	V	M	S	G	R	Site[c]	Source[d]
		Lentic		Lotic			
	V	M	S	G	R	Site[c]	Source[d]
Family Hydrophilidae							
Cercyon sp.	X	X	X			N	s
h *Helophorus* sp.				X	X	N	s
Order Trichoptera							
Family Hydropsychidae							
o *Cheumatopsyche* sp.				X	X	N,O	s
o *Hydropsyche betteni*				X	X	N	s
o *H. cuanis*				X	X	N	s
o *H. frisoni*				X	X	N,O	s
o *Hydropsyche* sp.				X	X	N,O	s
unident. spp.						N	s
Family Hydroptilidae							
Agraylea sp.	X					N	s
o *Mayatrichia* sp.	X	X	X	X	X	N,O	s
Ochrotrichia sp.	X	X	X	X	X	O	s
Oxyethira sp.	X			X	X	O	s
Family Leptoceridae							
h *Oecetis cinerascens*	X	X	X	X	X	O	s
h *Oecetis* spp.	X	X	X	X	X	O	s
Family Polycentropodidae							
c *Cernotina* sp.	X	X	X			O	s
o *Neureclipsis* spp.				X	X	N	s
Nyctiophylax moestus	X			X	X	O	s
Phylocentropus sp.		X	X			N,O	s
o *Polycentropus centralis*		X	X			O	s
o *P. remotus*				X	X	O	s
o *Polycentropus* sp.				X	X	N,O	s
Family Psychomidae							
unident. sp.							
Order Lepidoptera							
h unident. sp.						O	s
Family Cossidae							
Prionoxystus sp.	X					N	s
Order Diptera							
Suborder Nematocera							
Family Ceratopogonidae							
co *Palpomyia tibialis*	X	X	X			O	s
co *Palpomyia* sp.	X	X	X			O	s
c *Probezzia* sp.	X	X	X	X		N,O	s
c *Stilobezzia* sp.	X	X	X	X		O	s
Family Chaoboridae							
*c *Chaoborus punctipennis*	X					N	s
c *Chaoborus* sp.	X					N	s
Family Chironomidae							
o *Ablabesmyia* sp.	X	X		X		N,O	s
Anatopynia brunnea						O	s

TABLE 1.7. (*Continued*)

		Substrate[b]					
			Lentic		Lotic		
	Taxon[a]	V	M	S	G	R	Site[c]	Source[d]
	Family Chironomidae (*cont.*)							
do	*Brillia par*	X	X		X		O	s
	Chironomus sp.	X	X				N,O	s
	Cladotanytarsus sp.	X	X				N,O	s
c	*Clinotanypus* sp.	X	X	X			O	s
c	*Coleotanypus concinnus*	X	X	X			O	s
c	*C. scapularis*	X	X	X			O	s
c	*Coleotanypus* spp.	X	X	X			N,O	s
	Constempellina sp.				X		O	s
	Corynoneura sp.	X	X	X			N	s
hd	*Cricotopus bicinctus*	X	X	X	X		O	s
hd	*Cricotopus* spp.	X	X	X	X		N,O	s
c	*Cryptochironomus fluvus*	X	X				O	s
c	*C. parafluvus*	X	X				O	s
c	*C. ponderosus*	X	X				O	s
*o	*Cryptochironomus* nr. *digitatus*	X	X				O	a
*o	*Cryptochironomus* nr. *psittacinus*	X	X	X			O	a
	Cryptocladopelma sp.		X		X		O	s
	Cryptotendipes sp.		X		X		N,O	s
	Demicryptochironomus sp.						N	s
o	*Dicrotendipes modestus*	X	X				N,O	s
o	*D. nemodestus*	X	X				O	s
o	*Dicrotendipes* sp.	X	X				N,O	s
	Einfeldia sp.	X	X				N	s
	Endochironomus sp.	X					N	s
co	*Epoicocladius flavens*	X	X	X	X		O	s
co	*Epoicocladius* sp.	X	X	X	X		N,O	s
o	*Eukiefferiella* sp.	X	X	X	X		N,O	s
*ho	*Glyptotendipes senelis*	X	X	X			N	s
*ho	*Glyptotendipes* sp.	X	X	X			N,O	s
o	*Harnischia curtilamellata*	X					O	s
	Keifferulus sp.						N	s
	Lauterborniella sp.	X	X	X			N	s
	Micropsectra sp.	X	X				O	s
	Microtendipes sp.	X	X				O	s
	Nanocladius distinctus	X	X		X		O	s
	Nilothauma sp.		X	X			N	s
	Pagastiella ostansa	X	X	X			O	s
	Pagastiella sp.	X	X	X			O	s
co	*Parachironomus* spp.	X	X	X			N,O	s
o	*Paralauterborniella* sp.	X					N,O	s
	Paratanytarsus sp.	X	X	X	X		O	s
	Paratendipes sp.	X	X	X			N,O	s
co	*Pentaneura flavifrons*	X	X	X	X		O	s
co	*Pentaneura* sp.	X	X	X	X		O	s

(*continued*)

TABLE 1.7. (*Continued*)

Taxon[a]	V	M	S	G	R	Site[c]	Source[d]
	Substrate[b]						
	Lentic			Lotic			
Family Chironomidae (*cont.*)							
Phaenopsectra sp.	X	X	X			O	s
o *Polypedilum fallax*	X					O	a
o *P. haltare*	X					O	s
o *P. illinoense*	X					N,O	s
o *P. prob. illinosense*	X					O	a
o *P. ophioides*	X					O	s
h *Polypedilum* sp.	X					N,O	s
o *Potthastia longimanus*				X		O	s
o *Potthastia* sp.				X		N	s
Procladius spp.	X	X				N,O	s
Psectrocladius spp.	X	X				O	s
o *Rheotanytarsus* spp.				X		N,O	s
Stempellinella sp.	X	X	X	X		O	s
o *Stenochironomus* sp.	X					N,O	s
Stictochironomus flavicingula		X	X			O	s
Stictochironomus sp.		X	X			N,O	s
Tanypus sp.	X	X	X			N,O	s
*o *Tanytarsus* (= *Endochironomus?*) nr. *nigricans*	X			X	X	O	a
*o *Tanytarsus* spp.	X			X	X	N,O	s
Thienemanniella sp.	X	X	X	X	X	O	s
o *Thienemannimyia* gr.	X	X	X	X	X	O	s
Tribelos sp.	X	X				N,O	s
Trichocladius sp.			X			O	s
Xenochironomus sp.			X			O	s
unident. *Chironominae*	X	X		X	X	N,O	s
unident. *Diamesinae*	X	X		X	X	N	s
unident. *Orthocladinae*	X	X		X	X	N,O	s
unident. *Pentaneurini*	X	X		X	X	O	s
unident. spp.						O	b
Family Culicidae							
Aedes sp.		X	X			O	s
Family Simuliidae							
Ectemnia incenusta				X	X	N	s
Ectemnia sp.				X	X	N	s
Metacnephia sp.				X	X	N	s
Prosimilium ursinum				X	X	N,O	s
Simulium canadense				X	X	N	s
unident. spp.						O	s
Family Tipulidae							
Tipula sp.		X	X	X		N	s
unident sp.							
Suborder Brachycera							
Family Dolichopodidae							
Hydrophorus sp.	X	X	X			N	s

TABLE 1.7. (*Continued*)

Taxon[a]	V	Lentic		Lotic		Site[c]	Source[d]
		M	S	G	R		
Family Empididae							
Hemerodromia sp.		X	X	X	X	N	s
Family Stratiomyidae							
Stratiomyia sp.	X	X	X	X	X	O	s
Family Syrphidae							
Tubifera sp.	X	X	X	X		O	b
Phylum Bryozoa							
Class Ectoprocta							
Order Gymnolaemata							
Family Crisiidae							
unident. spp.	X					O	a
Family Nolellidae							
* *Nolella* sp.	X						
Order Phylactolaemata							
Family Lophopodidae							
Pectinatella magnifica	X					O	b
Family Plumatellidae							
*o *Plumatella repens*	X					O	a,b
Class Entoprocta							
Family Urnatellidae							
*o *Urnatella gracilis*	X					O	a
Phylum Chordata							
Subphylum Vertebrata							
Class Cephalaspidomorphi							
Order Petromyzontiformes							
Family Petromyzontidae							
Ichthyomyzon castaneus		X	X	X	X	S	t
Class Osteichthyes							
Order Acipenseriformes							
Family Polydontidae							
Polydon spathula	X	X	X	X	X	S	z
Order Lepisosteiformes							
Family Lepisosteidae							
Lepisosteus oculatus	X	X	X			E,S	t
c *L. osseus*	X	X	X	X		S	t
*c *L. productus* (= *oculatus?*)	X					O	a
L. spathula						E,S	t
Order Amiiformes							
Family Amiidae							
Amia calva	X	X				O,S	a,t
Order Elopiformes							
Family Elopidae							
Elops saurus	X	X	X			S	t
Order Anguilliformes							

(*continued*)

TABLE 1.7.   (*Continued*)

Taxon[a]	V	M	S	G	R	Site[c]	Source[d]
		Lentic		Lotic			
Family Anguillidae							
c *Anguilla rostrata*	X	X	X	X	X	O,S	a,t
Family Enchelidae							
Myrophis punctatus						O	a
Order Clupeiformes							
Family Clupeidae							
Brevoortia gunteri						S	t
o *Dorosoma cepedianum*	X	X	X	X	X	O,S	a,t
o *D. petenense*			X	X	X	O,S	a,b,t
Family Engraulididae							
Anchoa mitchilli		X	X			O,S	b,z
*o *A. mitchilli diaphana*		X	X			O	a
Order Cypriniformes							
Family Catostomidae							
o *Carpiodes carpio*	X	X				S	t
Cycleptus elongatus			X	X	X	S	t
Ictiobus bubalus			X	X		E,S	a,t
I. cyprinellus	X	X	X			S	z
Minytrema melanops	X	X				S	t
Moxostoma poecilurum		X	X	X		S	t
Family Cyprinidae							
Cyprinus carpio			X	X	X	O,S	b,t
o *Hybognathus nuchalis*		X	X			E	
*o *Notropis amnis*	X	X				E	a
N. atherinoides			X			O	a
N. emiliae	X	X				O	a
N. lutrensis			X	X		O	a
*o *N. roseus*	X		X			O	a
N. venustus		X	X			O,S	b,t
*o *Opsopoeodus emiliae*	X					O	a
*o *Pimephales vigilax perspicuous*	X	X				E	a
Order Siluriformes							
Family Ariidae							
Arius felis		X	X			S	t
Barge marinus						S	t
Family Ictaluridae							
Ictalurus furcatus			X	X	X	O,E,S	a,b,t
I. natalis	X	X	X	X		O,S	a,t
*o *I. punctatus*			X	X		O,S	a,b,t
Noturus gyrinus	X	X	X			O	a
Pilodictis olivaris		X	X			O,S	a,t
Order Percopsiformes							
Family Aphredoderidae							
Aphredoderus sayanus	X	X				O,S	a,t
Order Cyprinodontiformes							

TABLE 1.7. (*Continued*)

Taxon[a]	V	Lentic		Lotic		Site[c]	Source[d]
		M	S	G	R		
Family Belonidae							
Strongylura marina			X			O,S	b,t
Family Cyprinodontidae							
Fundulus chrysotus	X	X				E	b
F. grandis	X	X	X			O	b
F. olivaceus	X		X			O	a
Lucania parva	X	X				O	b
Family Poeciliidae							
*o *Gambusia affinis*	X	X				E	a
Poecilia latipinna	X	X				E	a
Order Atheriniformes							
Family Atherinidae							
Menidia beryllina			X			O	b
Order Syngnathiformes							
Family Syngnathidae							
Sygnathus scovelli	X	X				O	b
Order Perciformes							
Family Carangidae							
Caranx hippos	X		X			O	b
Family Centrarchidae							
* *Chaenobryttus coronarius*	X	X	X			O	a
*c *Elassoma zonatum*	X	X				O	a
Lepomis auritus			X			S	z
L. cyanellus	X	X	X			E	c
L. gulosus	X	X				S	t
L. humilis	X	X	X			S	t
*o *L. macrochirus*	X	X	X			O,S	a,b,t
*c *L. megalotis*	X	X	X	X	X	O,S	b,t
*c *L. microlophus*	X	X				O,S	b,t
L. punctatus	X	X				O	b
*o *L. punctatus miniatus*	X	X				O	a
Lepomis sp.	X	X	X	X	X	O	b
*o *Micropterus punctulatus*		X	X	X	X	O,S	b,t
M. salmoides	X	X	X	X		O,S	a,b,t
Pomoxis annularis	X	X				O,S	a,b,t
P. nigromaculatus	X		X			O,S	a,t
Family Gobiidae							
Gobionella shufeldti	X	X	X			O	b
Gobiosoma bosci						O	a
Microgobius gulosus	X	X				O	b
Family Mugilidae							
Mugil cephalus	X	X	X			O,S	a,b,t
Family Percidae							
c *Etheostoma asprigene*	X	X				O	a
c *Percina caprodes*			X	X	X	S	z

(*continued*)

TABLE 1.7. *(Continued)*

Taxon[a]	V	Lentic M	Lentic S	Lotic G	Lotic R	Site[c]	Source[d]
Family Percithyidae							
c *Morone chrysops*			X	X	X	S	z
M. mississippiensis			X	X	X	O,S	b,t
M. saxatilis			X	X	X	S	t
M. chrysops × *M. saxatilis*						S	z
M. mississippiensis × *M. saxatilis*						S	z
Family Sciaenidae							
c *Aplodinotus grunniens*		X	X			O,S	a,t
Cynoscion arenarius		X	X			S	t
Micropogonais undulatus		X	X			S	t
Family Sparidae							
Archosargus probatocephalus	X	X	X		X	S	t
Order Pleuronectiformes							
Family Bothidae							
Paralichthys lethostigma		X				O,S	b,t
Family Soleidae							
Trinectes maculatus		X	X			O	b
*o *T. maculatus fasciatus*	X	X	X			O	a

*Species found in the habitat described in the section "Structure of Aquatic Ecosystems."

[a] c, Carnivore; co, carnivore-omnivore; d, detritivore; do, detritivore-omnivore (devours organisms and all types of detritus); h, herbivore; hd, herbivore-detritivore; ho, herbivore-omnivore; o, omnivore.

[b] V, vegetation; M, mud; S, sand; G, gravel; R, rock.

[c] N, Lower Sabine River along Newton County, Texas; O, lower Sabine River along Orange County; S, lower Sabine River (unspecified); E, freshwater fishes recorded in brackish waters of the Sabine estuary.

[d] a, ANSP (1954) (spring 1953 only except for freshwater species recorded at estuarine sites); b, ANSP (1970) (summer); s, Sala, unpublished data; t, Seidensticker (1980); z, Seidensticker, personal communication.

were *Carteria cordiformis.* A few other flagellates were present that were primary producers. They were an unidentified species of *Chlamydomonas* and *Scourfieldia complanata.* Colonial algae that are primary producers in this habitat were *Eudorina illinoiensis, Gonium pectorale,* and *Pandorina morum.* Several euglenoids were found in and among the vegetation: *Anisonema acinus, Anisonema* sp., *Entosiphon ovatum, Notosolenus lens, Peranema deflexum,* and *P. trichophorum,* all primary producers. Other euglenas in this habitat were *Ascoglena* sp., *Euglena anabaena, E. fusca, E. haematodes, E. intermedia, E. viridis, Trachelomonas schauinslandii, T. verrucosa, T. volgensis,* and *T. volvocina,* all primary producers. Crawling over the debris in these vegetative habitats were the detritivore–omnivores *Arcella gibbosa* and *Arcella* sp. Also present were *Difflugia arcula, D. globosa,* and an unidentified species of *Difflugia.* Also present among the debris were the omnivores *Halteria grandinella* and *Strobilidium concicum.* Attached to the stems of some of the plants were colonies of *Uroleptus limnetis* as well as *Vorticella vestita. Uroleptus* is a bacterial detritivore, whereas the *Vorticella* is probably omnivorous.

Also omnivorous and attached to the stems of the plants were a sponge, *Spongilla* prob. *lacustris* and the detritivore-omnivore *Prostoma rubrum.* Rotifers found in this habitat were the omnivorous *Brachionus angularis, B. havanaensis, B. plicatilis, Keratella cochlearis,* and *Platyais patulus,* and the omnivore *Lecane luna.* Other omnivores found in this habitat were *Polyarthra remata* and *Filinia terminalis* and *Pedalia jenkinae.* Also found in this habitat was a gastropod, *Littoridina* sp.

Bivalves were found in this habitat associated with the debris of vegetation. They were *Rangia cuneata,* probably an omnivore; *Corbicula manilensis,* an omnivore; and *Eupera cubensis.* A few oligochaete worms, such as two unidentified species of the genus *Dero,* detritivore-omnivores, were also found here, as was a member of the family Naididae *Stylaria* sp., a detritivore-omnivore; an isopod, an unidentified species of the genus *Munna;* and the amphipod, *Gammarus fasciatus,* which tends to be omnivorous. A decapod, *Palaemonetes pugio,* was in and among the vegetative debris, as was *Penaeus setiferous,* both of which are omnivores, and *Rithropanopeus harrissi,* a carnivore.

Several insect larvae, the carnivores *Enallagma* near *signatum* and an unidentified species of the same genus, were in these vegetative habitats. The gerrid *Rheumatobates hungerfordi,* a carnivore, was crawling over the detritus. Also found in this habitat were the dipteran *Chaoborus punctipennis* and the chironomids *Cryptochironomus* near *digitatus* and *Cryptochironomus* near *psittacinus,* both omnivores. The omnivore-herbivores *Glyptotendipes senelis* and an unidentified species of the same genus were found in and among the debris, as were the chironomids *Tanytarsus* near *nigricans* and unidentified species of the same genus, all probably omnivores.

Attached to the stems of the aquatic plants were two ectoprocts, *Nolella* sp. and unidentified species belonging to the family Crisiidae. Also present were the bryozoan *Plumatella repens,* an omnivore, and *Urnatella gracilis,* which is probably omnivorous. Several fish were found in and among the roots of the emergent vegetation and swimming over the debris. They were the carnivore *Lepisosteus productus;* the cyprinid *Notropis amnis,* perhaps an omnivore with herbivorous preferences; *N. roseus,* an omnivore; *Opsopoeodus emiliae,* an omnivore; and *Pimephales vigilax* var. *perspicuous.* Other fish that were present in this habitat were *Gambusia affinis,* an omnivore; *Chaenobryttus coronarius;* the carnivorous *Elassoma zonatum; Lepomis macrochirus,* an omnivore; *L. megalotis,* a carnivore; *L. microlophus,* a carnivore; *L. punctatus miniatus,* an omnivore; and *Trinectes maculatus fasciatus,* an omnivore.

Thus we see in the vegetative habitat, species representing the various stages of nutrient and energy transfer. At each stage of nutrient and energy transfer—that is, primary producers, detritivores, detritivore-omnivores, herbivore-omnivores, omni-vores, and carnivores—there are many species representing many different major groups of organisms.

Lentic Habitats.   Lentic habitats occur in pools, along the sides of the river chan-nel away from the main force of the flow, in and among rooted aquatics, and some-times behind obstructions such as logs that inhibit the flow of the river. In these lentic habitats were several dicotyledons: *Ceratophyllum demersum,* several unidentified species of the genus *Polygonum,* and *Hydrocotyl* sp. Also present were various monocotyledons. These were usually partially submergent and partially emergent plants, such as unidentified species of *Sagittaria.* Floating on the water with partially submerged stems were *Vallisneria americana* and an unidentified species of the genus *Potamogeton.*

The blue-green algae *Anabaena tortulosa, Lyngbya sordida, L. lutea, Phormidium retzii,* and *P. subuliforme* were present attached to the stems of the rooted aquatics. Also present was the desmid *Closterium moniliferum.* Several diatoms were present in this habitat: *Melosira distans* var. *alpigena, Synedra ulna,* and an unidentified species of the genus *Synedra.* Also present in these lentic habitats were *Achnanthes temperei,* specimens of which were attached to the detritus in the bottom of pools. Also present were *Gomphonema gracile* var. *naviculoides* and *G. parvulum. Frustu-lia rhomboides* and *Frustulia rhomboides* var. *crassinervia* were in and among the detritus in these lentic areas, as were *F. weinholdii* and *Gyrosigma scalproides. Nav-icula confervacea* was common along the edges of the river in these lentic habitats. *N. crucicula, N. mobiliensis, N. mutica, N. schroeteri, N. subatomoides,* and some unidentified species of the genus *Navicula* were present on the surface of the mud in these lentic habitats. In shallow pools were *Pinnularia acrosphaeria* and *Stauroneis crucicula.* Other diatoms found in these lentic habitats were several species of the genus *Nitzschia: N. clausii, N. filiformis, N. keutzingiana, N. sigma,* and *N. try-blionella* var. *victoriae. Surirella striatula* was found in the same habitats as the various species of *Nitzschia.* In the mud were *Ceratophyllum demersum* and *Poly-gonum* spp.

Various protozoans were present in these lentic habitats: the primary producers *Cryptomonas marssonii* and *Cyathomonas truncata.* Other protozoans present were *Carteria cordiformis,* a primary producer; the euglena, *Notosolenus lens,* a primary producer, and *Peranema deflexum* and *P. trichophorum,* also primary producers. Other protozoans found in this habitat were *Lepocinclis texta,* a primary producer, as well as *Phacus curvicauda* and *P. orbicularis.* In pool-like areas were the primary producers *Trachelomonas schauinslandii, T. verrucosa, T. volgensis,* and *T. volvocina.* Various amoeba associated with the muddy sand habitat were *Arcella gibbosa, Arcella* sp., *Difflugia arcula, D. globosa,* and *Difflugia* sp. These are detritivore-omnivores.

Several ciliates were in and among the detritus in these lentic habitats. They were an unidentified species of the genus *Lionotus, Cinetochilum margaritaceum, Cyr-tolophosis mucicola,* and *Frontonia leucas.* These are bacterial and diatom feeders. Other protozoans that live largely on bacteria that were found in and among the detritus in these lentic habitats were *Cyclidium glaucoma* and *C. heptatrichum.* The hypotrichs *Aspidisca costata* and *A. lynceus* were found among the detritus in pools and along the margins of the river. These habitats were also favorable for four taxons belonging to the genus *Euplotes,* which are omnivores (Table 1.7). Also present in this habitat was *Keronopsis spectabilis,* which is probably an omnivore. Found here was the sponge, *Spongilla c.f. lacustra.*

A mollusc found in this habitat was *Littoridina* sp., which belongs to the family Amnicolidae. An oligochaeta, *Stylaria* sp., which is probably a detritivore-omnivore, was in and among the detritus. The amphipod *Corophium lacustre* was collected in and among the detritus, as was *Gamarus fasciatus,* which is an omnivore. The crab, *Uca pugnax,* which is an omnivore, was found in pools. Here were also *Palaemonetes pugio, P. paludosus,* and *P. vulgaris,* omnivores, and *Penaeus setiferous,* an omnivore. *Callinectes sapidus acutidens,* a carnivore, was also found here, as was *Rithropanopeus harrissi.*

On the surface of debris, particularly on the stems of plants that are living in these lentic habitats, were several odonates: *Enallagma* near *signatum,* a carnivore, and *Nasiaeschna* sp., a carnivore. The chironomids *Cryptochironomus* near *digitatus* and *C.* near *psittacinus* were found here. Both of these species are probably omnivorous, feeding on bits of algae, detritus, bacteria, and protozoans in these lentic habitats. Also present were two taxons belonging to the genus *Glyptotendipes,* omnivores with a tendency toward being herbivorous. They were found in the sandy mud.

A few fish were collected in these lentic habitats. They were omnivorous *Anchoa mitchilli diaphana* and *Notropis amnis.* Also present were the omnivorous *Notropis roseus* and *Pimephales vigilax perspicuous.* They were present in the pools, which were often near the edges of the river. In one of the deeper pools was the catfish, *Ictalurus punctatus,* an omnivore. In and among the vegetation in these lentic habitats were the omnivorous *Gambusia affinis* and the centrarchid *Chaenobryttus coronarius.* Other fish in these lentic habitats were the carnivorous *Elassoma zonatum;* the omnivorous *Lepomis macrochirus;* the carnivorous *L. megalotis;* the carnivorous *L. microlophus;* the omnivorous *L. punctatus miniatus;* and *Micropterus punctulatus.* Also present was *Trinectes maculates fasciatus,* an omnivore.

As can be seen from the species present, this lower station of the Sabine River is not a typical freshwater large river habitat because of the saline influence of the nearby estuary.

Lotic Habitats. Lotic habitats existed around snags and debris caught in the bed of the river. These were excellent habitats for a fairly large variety of aquatic life. On the surface of snags was the blue-green alga, *Anabaena tortulosa.* Also present were *Lyngbya lutea* and the blue-green algae *Microcoleus tenerrimus* and *Phormidium retzii.* Attached to debris in these lotic habitats were *Ulothrix implexa* and filaments of *Oedogonium* spp. and *Rhizoclonium hieroglyphicum.* Here and there in areas where the current was weaker were filaments of *Spirogyra* spp.

These lotic habitats were very favorable for several species of diatoms. Attached to debris where the current was not very strong was *Melosira distans* var. *alpigena.* Also present in areas where the current was not very strong was *Terpsinoe musica.*

Synedra goulardi and *S. ulna* were found in these lotic habitats where the current was moderate in swiftness. Here one also found *Eunotia formica* among the gravel in lotic habitats with moderately fast current. Also here was *Eunotia praerupta* var. *bidens*. Attached to the surface of debris were *Achnanthes* sp., *Cocconeis* sp., *Amphora* sp., and *Cymbella tumida*. A few Gomphonemas were found in this habitat. They belonged to the species *Gomphonema parvulum*. On the surface of the muddy sand in these lotic habitats was *Caloneis bacillum*. Also present in areas where the current was moderate to slow were *Frustulia rhomboides*, *F. rhomboides* var. *crassinervia*, and *Gyrosigma scalproides*. A few species of *Navicula* were found in the sandy substrates. They were *Navicula confervacea*, *N. mutica*, and *N. schroeteri*. *Rhopalodia gibberula* was found attached to the debris in these lotic habitats. In the surface of the sandy mud were *Nitzschia brevissima*, *N. clausii*, *N. fasciculata*, *N. filiformis*, and *N. fonticola*. Other species found on the surface of the sandy mud were *N. keutzingiana*, *N. lorenziana* var. *subtilis*, *N. obtusa*, *N. tryblionella* var. *levidensis*, and *N. tryblionella* var. *victoriae*.

Attached to debris in moderately swift water was the red alga, *Compsopogon coeruleus*. In relatively protected areas associated with the gravel in these lotic habitats were *Paramecium multimicronucleatum*, an omnivore, and *Halteria grandinella*, an omnivore. Also found in this habitat was *Strobilidium concicum*. *Uroleptus limnetis* was also in this habitat. Several vorticellids of the species *Vorticella vestita* were present. Attached to debris was the sponge, *Spongilla* prob. *lacustris*. Crawling over the debris was a mollusc, *Littoridina* sp., an omnivore. Also present was *Mytilopsis leucophaeatus*. This unionid is a filter feeder. Atached to the debris was the barnacle, *Balanus amphitrite niveus*, a filter feeder and an omnivore. An omnivorous amphipod, *Corophium lacustre*, was also found in these lotic habitats in the gravel. Here also was the gammarid, *Gammarus fasciatus*, an omnivore. *Palaemonetes pugio*, an omnivore, was also in these lotic habitats, as was *Rithropanopeus harrissi*, a carnivore.

The anisopteran *Nasiaeschna* sp., which is mainly a detritivore, possibly an omnivore, was found here in the lotic habitats. A gerrid, *Rheumatobates hungerfordi*, was present. A few chironomids were in and among the sandy mud: *Tanytarsus* nr. *nigricans*, an omnivore, and unidentified species of *Tanytarsus*, probably also omnivorous.

A few fish were found in these lotic habitats. They were the carnivorous *Lepomis megalotis* and *Micropterus punctulatus*, an omnivore with carnivorous tendencies. Also present was *Ictalurus punctatus*, an omnivore.

SUMMARY

The Sabine River is in the southwestern United States and flows southeast and south for 180 mi from Hunt County in eastern Texas, to Sabine Lake, through Sabine Pass, into the Gulf of Mexico. The structures of the aquatic ecosystems in a freshwater reach near the source of the river and in a freshwater site influenced by salt water in the upper part of the estuary are described. Each of these communities supports a large number of species, which function in varying ways to secure the breakdown of waste and the transfer of nutrients and energy through the food web. In both the upper Sabine and lower Sabine areas, the base of the food web is, on the one hand, algae and protozoans that are photosynthetic and, on the other hand, protozoans and other organisms that feed upon detritus and bacteria. These ecosystems include species that are primarily herbivores and detrivores, omnivores, and carnivores. At each stage of nutrient and energy transfer there are many species representing different phylogenetic groups. The four stages of nutrient and energy transfer function in and among the vegetation, in pools, and in lotic areas where current is present, such as around snags and debris in the lower Sabine River and in association with rocks and gravel and snags in the upper reaches of the river. In each area we find many different habitats occupied by many species functioning in the transfer of nutrients and energy through the ecosystem or food web. In the upper station all the species are characteristic of fresh water of low or moderately high conductivity, whereas in the lower study area, freshwater species as well as those that can tolerate slightly brackish-water conditions are present.

BIBLIOGRAPHY

Academy of Natural Sciences of Philadelphia. 1954. Sabine River, Texas, Summer 1952–Spring 1953. Stream survey report for the Sabine River Plant—Polychemicals Department, E.I. du Pont de Nemours and Company. ANSP, Philadelphia, Pa. 116 pp.

Academy of Natural Sciences of Philadelphia. 1963. Sabine River biological survey 1962 for the E.I. du Pont de Nemours and Company Sabine River Plant. ANSP, Philadelphia, Pa. 80 pp.

Academy of Natural Sciences of Philadelphia. 1970. Sabine River survey, 1969. River survey report for the Sabine River Works, E.I. du Pont de Nemours and Co. ANSP, Philadelphia, Pa. 117 pp.

Academy of Natural Sciences of Philadelphia. 1996. 1995 Sabine River Studies for Texas Eastman Company. Rep. 96–9. 193 pp.

Benke, A. C., and J. L. Meyer. 1988. Structure and function of a blackwater river in the southeastern U.S.A. Int. Verein. Limnol. Verh. 23(1): 1209–1218.

Blair, W. F. 1950. The biotic provinces of Texas. Tex. J. Sci. 2(1): 93–117.

Fenneman, N. M. 1938. Physiography of the eastern United States. McGraw-Hill, New York. 714 pp.

Hubbs, C. 1957. Distribution patterns of Texas fresh-water fishes. Southwest. Nat. 2(2-3): 89–104.

Hughes, L. S., and D. K. Leifeste. 1965. Reconnaissance of the chemical quality of surface waters of the Sabine River basin, Texas and Louisiana. Water-Supply Pap. 1809-H. U.S. Geological Survey, Washington, D.C.

Lee, D. C., C. R. Gilbert, C. H. Hocutt, R. E. Jenkins, D. E. McAllister, and J.R. Stauffer, Jr. 1980. Atlas of North American freshwater fishes. North Carolina State Museum of Natural History, Raleigh, N.C. 854 pp.

Respess, R. O. 1985. Intensive survey of Sabine River Segment 0505, September 24–28, 1984: hydrology, field measurements and water chemistry. IS 85-01. Texas Water Commission, Austin, Texas. 43 pp.

Sabine River Authority of Texas. 1992. Regional assessment of water quality, Sabine River Basin, Texas. 1.0-1-2.28-14. Texas Natural Resources Conservation Comission, Austin, Texas.

Sabine River Authority of Texas. 1994. Regional assessent of water quality. Texas Natural Resources Conservation Commission, Austin, Texas. 152 pp.

Sabine River Authority of Texas. 1996. Regional assessment of water quality. Texas Natural Resources Conservation Commission, Austin, Texas. 142 pp.

Sala, G. M., J. W. Tatum, and D. S. Parsons. 1979. Gulf Coast Division water quality monitoring program, lower Sabine River watershed and canal system, February, 1972–December, 1977. Tech. Rep. 79-1. Sabine River Authority of Texas, Texas Natural Resources Conservation Commission, Austin, Texas.

Seidensticker, P. Personal communication. Texas Parks and Wildlife Department, Austin, Texas.

Seidensticker, E. P. 1980. Existing reservoir and stream management recommendations: Sabine River. Performance Report for Job A, Dingel–Johnson Project F-30-R-5. Texas Parks and Wildlife Department, Austin, Texas. 30 pp.

Twidwell, S. R. 1978. Intensive surface water monitoring survey for segment 0505, Sabine River, Toledo Bend headwater to U.S. 271 near Gladwater. Prepared by Texas Water Quality Board, Austin, Texas. 49 pp.

U.S. Geological Survey. 1975a. Surface water supply of the United States, 1966–1970. Part 8. Western Gulf of Mexico basins. Vol. 1. Basins from Mermentau River to Colorado River. Water-Supply Pap. 2122. USGS, Washington, D.C.

U.S. Geological Survey. 1975b. Quality of surface waters of the United States, 1970. Part 8. Western Gulf of Mexico basins. Water-Supply Pap. 2157. USGS, Washington, D.C.

CHAPTER 2

Guadalupe River

INTRODUCTION

The Guadalupe River arises in Kerr County, south-central Texas, and flows 712 km in a southeasterly direction, until it empties into the Gulf of Mexico at San Antonio Bay (Epperson and Short, 1987) (Figure 2.1). For the first 260 km, the Guadalupe flows through the Edwards Plateau. (Hereafter, this reach will be referred to as the upper Guadalupe.) This plateau, which is the southeastern extension of the High Plains Physiographic Province, is characterized by highly dissected karst topography and clear, rocky, spring-fed streams. Geologically, the Edwards Plateau is dominated by Lower Cretaceous limestone, and it is underlain by major groundwater reservoirs [Epperson and Short, 1987; Fenneman, 1931; Hubbs et al., 1953; U.S. Geological Survey (USGS), 1970].

The Guadalupe leaves the Edwards Plateau as it descends the Balcones Escarpment and enters the Texas extension of the Coastal Plain Physiographic Province. Shortly after entering the coastal lowlands, the river is joined by the Comal River, which is essentially an outlet for Comal Springs, a major network of springs created by the upward diversion of Edwards Plateau aquifers at the Balcones Fault Zone near the town of New Braunfels. From here the Guadalupe flows through rolling to flat topography for approximately 170 km (hereafter referred to as the middle Guadalupe) until it is joined by its major tributary, the San Marcos River. This tributary also receives a substantial proportion of its flow from groundwater issuing from the Balcones Fault Zone (at San Marcos Springs), but most of the tributaries of the San Marcos (e.g., Blanco River and Plum Creek) are subject to intermittency. Following

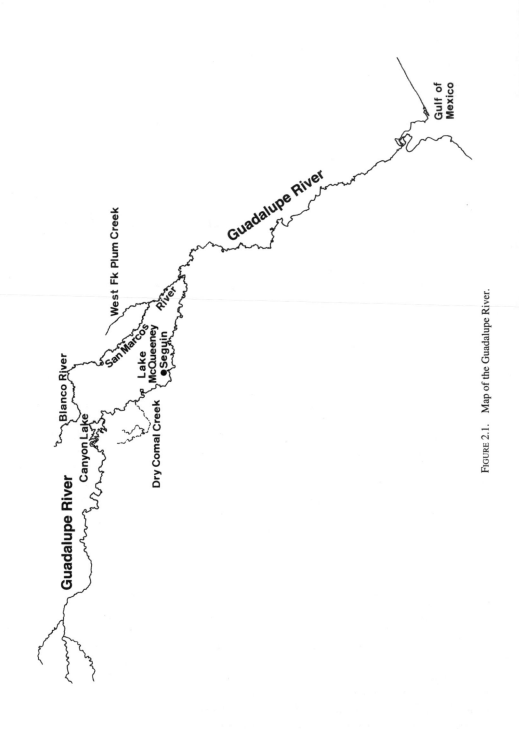

FIGURE 2.1.   Map of the Guadalupe River.

the confluence with the San Marcos, the Guadalupe River is joined by the Peach, Sandies, and Coleto creeks as it flows over the progressively flatter topography of the lower Coastal Plain. It is joined by the San Antonio River near San Antonio Bay (Epperson and Short, 1987; Fenneman, 1938; Hubbs et al., 1953; USGS, 1970; Young and Bayer, 1979).

Blair (1950) designated the Edwards Plateau as the Balconian Biotic Province, which is characterized by semiarid to subhumid climate and shallow, rocky soils. Climax watershed vegetation is dominated by scrub forests of Mexican juniper (*Juniperus ashei*), Texas oak (*Quercus shumardi* v. *texana*), and live oak (*Q. virginiana*). The Coastal Plains portion of the watershed belongs to the Texas Biotic Province, which is characterized by a moist, subhumid climate and deep soils of clay, loam, and sand. The climax vegetation is an oak–hickory forest dominated by post oak (*Q. stellata*), blackjack oak (*Q. marilandica*), and hickory (*Carya texana*). Floodplain and riparian assemblages, characterized by bald cypress (*Taxodium distichum*), pecan (*Carya illioensis*), elm (*Ulmus crassifolia*), hackberry (*Celtis laevigata*), and *Q. virginiana*, are common to both regions (Young and Bayer, 1979).

Rugged topography, dry climate, and poor soils have generally limited land use in the Edwards Plateau to grazing and recreation. In contrast, agricultural development of the lowland Texas Biotic Province has been extensive, with cattle production and cotton accounting for most of the land use. In addition, petroleum and natural gas extraction have in the past affected much of the San Marcos subdrainage and the mainstem Guadalupe. Urban and industrial development is concentrated in the lower Guadalupe near Victoria [Academy of Natural Sciences of Philadelphia (ANSP), 1953; USGS, 1970; Young and Bayer, 1979].

PHYSICAL CHARACTERISTICS

The portion of the Guadalupe River considered for this chapter is a middle Guadalupe (upper Coastal Plain) site located near the town of Seguin. This site, which is located approximately 38 km downstream from the confluence of the Guadalupe and the Comal River, was studied by the Academy of Natural Sciences from 1949 through 1973 (Figure 2.1). Bankfull width at the Seguin site was approximately 45 m. The banks were steep and the shallow littoral zones usually extended for only 1 m. Dropoff was typically rapid, and most of the channel was relatively deep (averaging 2.4 m during late summer). The site also included a moderately sized, tree-covered island which was separated from the shore by a shallow side channel (ANSP, 1949, 1963). Riparian vegetation, dominated by *Taxodium distichum* and *Carya illioensis*, was well developed (ANSP, 1963, 1967). Substrates included mud, marl, gravel, large cobble, and woody debris. Mud was a very common substrate in late summer surveys conducted in 1949, 1950, 1952, 1962, and 1966 (ANSP, 1949, 1951, 1953, 1963, 1967), but hurricane-associated storm flows (1972) scoured most of this mud from the channel by the time the site was surveyed in the summer of 1973 (ANSP, 1974).

Prior to the completion of Canyon Reservoir in 1974, the hydrology of the Guadalupe River was determined by the interaction of highly variable surface runoff and stable groundwater inflow. Precipitation, which occurs almost exclusively as rain, averages 60 to 75 cm yr⁻¹ in the Edwards Plateau and 75 to 100 cm yr⁻¹ in the Coastal Plain. On average, precipitation is relatively uniform throughout the year, but the month-to-month and year-to-year variability is high (Visher, 1954). This variability is reflected in the discharge history of the Blanco River, which is primarily dependent on surface runoff. For instance, flow in the Blanco near Kyle was continuous from October 1959 until mid-July 1963, but the stream was dewatered from mid-July 1963 through January 1964; flow for the remainder of 1964 was intermittent. In contrast, flows from the Comal and San Marcos springs are relatively constant, averaging 8.0 and 4.4 m³ s⁻¹, respectively (USGS, 1961, 1970).

The interaction between surface runoff and groundwater inflow is evident in the comparisons of discharge in the Guadalupe upstream from the Comal River and in the Comal River (Table 2.1). Flow recorded at the Guadalupe River site above the Comal River results from the combination of surface runoff and numerous small springs within the Edwards Plateau. This results in a continuous but variable-flow regime. This contrasts with the nearly uniform flow regime of the Comal River, which with the exception of occasional storm flows is derived almost exclusively from the Comal

TABLE 2.1.   *Mean Monthly Discharge (m³ s⁻¹) in the Unregulated (pre-Canyon Reservoir) Guadalupe River, Water Years 1960, 1961, and 1962*

	Guadalupe River Above Comal River			Comal River		
	1960	1961	1962	1960	1961	1962
Oct.	42.4	51.6	4.5	9.2	12.1	9.8
Nov.	9.3	28.9	5.4	8.8	9.8	10.0
Dec.	9.2	32.4	5.0	8.9	10.2	9.3
Jan.	13.2	29.8	4.4	9.2	10.3	9.1
Feb.	12.8	50.5	4.1	9.3	10.5	8.9
Mar.	9.9	27.8	3.8	9.3	10.3	8.6
Apr.	10.5	16.0	4.4	9.4	9.8	8.5
May	8.4	9.4	3.7	9.0	9.1	8.1
June	8.5	16.5	5.4	9.5	9.2	7.6
July	8.0	10.3	1.5	8.4	10.3	6.7
Aug.	27.8	6.5	0.5[a]	8.5	9.2	5.7
Sept.	9.2	4.8	2.4[a]	8.5	8.9	6.3
Annual mean[b]	14.2	23.6	3.7	9.0	10.0	8.2
Maximum	625.8	368.1	32.3	33.4	54.1	26.1
Minimum	2.6	3.8	0.2[c]	7.7	8.3	5.5

Source: Data from water year 1960 modified from USGS (1961); data from water years 1961 and 1962 modified from USGS (1970). Data converted from ft³/s to m³ s⁻¹.

[a]Flow was diverted in August and September 1962 while the deepwater impoundment, Canyon Reservoir, was operated as a detention basin (August 1962 through June 1964). Canyon Reservoir became operational in 1964.

[b]The 37-year mean (1928–1965) at the Guadalupe above Comal River is 10.0 m³ s⁻¹. The 33-year mean (1932–1965) for the Comal River is 8.0 m³ s⁻¹.

[c]Minimum discharge (water year 1962) exclusive of August and September diverted flows.

Springs. Flow from these two rivers accounts for most of the discharge in the Guadalupe at the Academy's survey site near Seguin, where the studies of aquatic life were made (Hannan et al., 1973; USGS, 1961, 1970; Young and Bayer, 1979).

Hydrology in the Guadalupe is also affected by main-stem impoundments. Two series of small, surface release dams are located in the Guadalupe. One series is located within the Edwards Plateau (upper Guadalupe) (Stanley et al., 1990). The second series consists of six small hydroelectric dams located in the middle Guadalupe between the confluence with the Comal River and that of the San Marcos River. Three of these dams, including Lake Dunlap and Lake McQueeny, are located upstream from the site at Seguin. In general, flow variations from hydroelectric releases are relatively modest (0.3 m) (Young et al., 1972). On the other hand, the Guadalupe's hydrology is substantially influenced by Canyon Reservoir, a deep-storage 3300-ha impoundment located approximately 30 km upstream from the confluence with the Comal River (Hannan and Broz, 1976) (Table 2.2). Flow in the Guadalupe has been regulated since the completion of the dam at Canyon Reservoir in 1974 (USGS, 1970).

Turbidity is relatively high in the Guadalupe at the Academy's survey site near Seguin. It ranged from 54 to 140 ppm during late summer (ANSP, 1953, 1963). Similarly, Young et al. (1972) reported that turbidity at another site near Seguin (station 7) generally ranged between 40 and 100 JTU, while Secchi disk transparencies generally ranged between 20 and 40 cm. As would be expected, turbidities were positively associated with discharge. Much of this turbidity is incurred as the Guadalupe

TABLE 2.2.   *Mean Monthly Discharge m^{3}s^{-1}) from Canyon Reservoir, Comal River, and Lake Dunlap (Upper Coastal Plain), Guadalupe River, Texas, February 1969 to January 1970*

	Canyon Reservoir	Comal River	Lake Dunlap[a]
Feb.	6.0	8.8	17.4
Mar.	6.0	9.2	17.8
Apr.	6.0	9.1	19.4
May	10.8	9.8	26.5
June	9.7	8.9	22.2
July	3.9	7.5	13.2
Aug.	1.8	6.2	10.6
Sept.	3.1	7.3	11.8
Oct.	18.8	7.6	28.4
Nov.	15.1	8.1	24.2
Dec.	11.5	8.6	23.9
Jan.	21.4	8.7	31.8
Annual mean	9.5	8.3	20.6

Source: Modified from Hannan et al. (1973).
[a]Lake Dunlap is approximately 17 km upstream from the Academy survey site near Seguin, Texas.

passes through the Coastal Plain, probably because of crop and livestock production. Hubbs et al. (1953) reported that the upper Guadalupe was clear except immediately after major rainstorms, while turbidity in the Comal River averaged only 2 JTU (Hannan and Broz, 1976). Turbidity increases within the middle Guadalupe were demonstrated by the study by Young et al. (1972). Turbidities in the first 10 km below the entry of the Comal River never exceeded 40 JTU, while the turbidities recorded at most downstream sites usually exceeded 60 JTU throughout the year and commonly exceeded 100 JTU during April and May. Significantly, the smaller hydroelectric impoundments in this reach were ineffective in reducing turbidities.

Stream temperatures in the Guadalupe River are consistently warm. Young et al. (1972) reported that temperatures from their station near Seguin (station 7) generally ranged between 15 and 30°C. Surface temperatures at the Academy's survey site near Seguin during late summer ranged from 25.6 to 31.0°C (ANSP, 1949, 1951, 1953, 1963, 1974).

CHEMICAL CHARACTERISTICS

The Guadalupe River is a very hard, alkaline stream dominated by calcium and magnesium hardness and bicarbonate alkalinity. For example, late summer values for total hardness ranged from 188 to 245 ppm at the Academy's survey site near Seguin during 1949, 1950, 1952, and 1962. Dissolved calcium usually exceeded 60 ppm, while dissolved magnesium ranged from 8 to 33 ppm. Sulfate and chloride typically ranged from 20 to 30 ppm and 15 to 25 ppm, respectively. Bicarbonate alkalinity typically exceeded 200 ppm, and pH ranged from 7.7 to 8.0 (ANSP, 1949, 1951, 1953, 1963, 1974) (Table 2.3).

Chemical characteristics at another site near Seguin (station 7) exhibited similar characteristics throughout the year. Bicarbonate alkalinity usually ranged between 180 and 220 mg L^{-1}, while carbonate alkalinity was undetectable. pH ranged from 7.0 to 7.6 and specific conductance typically ranged from 480 to 520 μS cm^{-1}. Seasonally, specific conductance was generally highest during low flows and lowest during high flows. In contrast, seasonal patterns in alkalinity and pH were more strongly associated with seasonal patterns in primary production. The pH maximum coincided with the late winter/early spring maximum in water column chlorophyll *a* concentrations, while minima occurred for midsummer through autumn, when chlorophyll *a* concentrations were low. Alkalinity differed from pH in that it appeared to be more strongly affected by upstream primary production than by local planktonic production. Alkalinity near Seguin was lowest during the same period that exhibited maximum dissolved oxygen and chlorophyll *a* concentrations in two upstream impoundments, Lake Dunlap and Lake McQueeny (Young et al., 1972).

Late summer macronutrient concentrations at the Academy's site near Seguin were moderately high and suggestive of enriched conditions. Nitrate was the primary form of inorganic nitrogen during the 1949, 1962, and 1973 surveys, while ammonia was prevalent during the 1950 and 1952 surveys. PO_4-P, ranging from 0.004 to 0.011 ppm, was moderately low. The ratio of total inorganic nitrogen to inorganic phosphorus ranged from 42 to 105: 1 during the 1949, 1950, and 1952 surveys (ANSP,

TABLE 2.3. *Low-Flow Chemical Characteristics of the Guadalupe River near Seguin, Texas*[a]

	Sept. 1949	Sept. 1950	Sept. 1952	Aug. 1962	Aug. 1973
Inorganic constituents					
Total hardness	188	232	245	208	205
Calcium hardness	155	152	—	74	165
Magnesium hardness	33	80	—	134	42
Calcium	62	61	64	30	65
Magnesium	8	19	21	33	10
Sulfate	20	20	20	30	20
Chloride	15	20	18	25	18
Specific conductance	—	368	350	500	380
Alkalinity					
Carbonate alkalinity	3	16	—	7	0
Bicarbonate alkalinity	211	208	202	220	182
pH	7.8	7.9	7.7	7.8	7.7
Nutrients					
NO_3-N	0.531	0.036	0.070	0.214	0.55
NO_2-N	0.011	0.019	0.020	0.004	0.011
NH_3-N	0.042	0.153	0.330	<0.002	0.04
PO_4-P	0.011	0.005	0.004	0.039	—
Silica	11	12	14	16	11
Dissolved oxygen					
Concentration	5.2	5.8	4.7	4.8	6.9
Percent saturation[b]	66	72	60	62	81

Source: 1949 data from ANSP (1949); 1950 data from ANSP (1953); 1952 data from ANSP (1953); 1962 data from ANSP (1963); 1973 data from ANSP (1974).

[a]Except for pH, specific conductance (μS) and percent saturation, data from 1949, 1950, 1952, and 1963 are expressed as ppm, while data from 1973 are expressed as mg L⁻¹.

[b]Percent oxygen saturation inferred from concentration and temperature.

1949, 1951, 1953), indicating phosphorus-limiting conditions. However, the 1962 ratio (5.6:1) suggested nitrogen-limiting conditions (ANSP, 1963). Silica concentrations were consistently high (Table 2.3).

Seasonal and longitudinal investigations of the middle Guadalupe indicated that both local and upstream processes significantly influence macronutrient dynamics. Inorganic nitrogen entering this reach is most strongly influenced by the relative contributions of the upper Guadalupe (via Canyon Reservoir) and the Comal River to discharge. Releases from Canyon Reservoir are characterized by relatively low concentrations of nitrate-nitrogen (0.01 to 0.60 mg L⁻¹), largely because of autotrophic uptake within the impoundment. On the other hand, nitrate-nitrogen in the Comal River is consistently high (1.2 to 1.6 mg L⁻¹ presumably because of groundwater contamination within the Edwards Plateau (Hannan and Broz, 1976; Hannan et al., 1973). The influence of these two sources on the concentrations of nitrate-nitrogen was evident at a site just downstream from their confluence sampled during 1969. Nitrate-nitrogen concentrations were less than 0.08 mg L⁻¹ during October and

November, when flow was strongly dominated by releases from Canyon Reservoir. On the other hand, concentrations ranged between 0.08 and 0.12 mg L^{-1} when inputs from the Comal River were either greater than or comparable to inputs from Canyon Reservoir (Hannan et al., 1973).

Seasonal maxima in nitrate-nitrogen (>0.12 mg L^{-1}) throughout the middle Guadalupe occurred during February, March, April, and November and were associated with high surface runoff. Surface runoff was also associated with most maxima in nitrite-nitrogen (>0.01 mg L^{-1}) and ammonia-nitrogen (<0.10 mg L^{-1}), but summer maxima were also recorded downstream from point sources at New Braunfels and Seguin (below the Academy's site and station 7) and in Lake Gonzales (approximately 70 km downstream from Seguin) (Hannan et al., 1973).

Seasonal minima in nitrate-nitrogen (<0.08 mg L^{-1}) were recorded from June through October. Minima recorded during October and November are attributable to predominance of Canyon Reservoir releases. Significantly, these low October and November concentrations persisted throughout the middle Guadalupe. In contrast, the minima recorded from June through September were limited to two reaches. One reach, which included the sites near Seguin, extended for 20 to 40 km below Lake Dunlap and Lake McQueeny. The summer nitrate-nitrogen minima within this reach were strongly associated with high phytoplankton and macrophyte biomass as well as high daytime dissolved oxygen content within the two lakes. Presumably, downstream minima are the result of autotrophic uptake of nitrogen within these two lakes. The second set of summer minima were located in the vicinity of Lake Gonzales and Wood Lake, presumably because of high macrophyte uptake (Hannan et al., 1973).

Sewage treatment outfalls at New Braunfels and Seguin are major sources of phosphorus in the middle Guadalupe. Total phosphorus concentrations or phosphate in streamwater entering from the upper Guadalupe was 0.05 mg L^{-1}, while those entering from the Comal River were probably less than 0.02 mg L^{-1}. In contrast, concentrations recorded below the two outfalls commonly exceeded 0.10 mg L^{-1}. Downstream declines in total and inorganic phosphate phosphorus were recorded throughout the year, but the declines were most pronounced during periods of high monthly discharge. Concentrations of total phosphate phosphorus at station 7 (a short distance upstream from the Seguin outfall) ranged between 0.02 and 0.08 mg L^{-1}, with maxima occurring during February and November. Minima occurred in spring and summer and are associated with high autotrophic production in Lake Dunlap and Lake McQueeny (Hannan et al., 1973).

Warm temperatures and moderate enrichment typically result in relatively low levels of dissolved oxygen. For example, late summer dissolved oxygen concentrations at the Academy's site near Seguin ranged from 4.7 to 5.8 ppm (60 to 72% saturation) for the 1949 to 1962 surveys (ANSP, 1949, 1951, 1953, 1963). The river was apparently more highly oxygenated during the 1973 survey, possibly because of increased flow or the hurricane-associated scouring of oxygen-demanding muddy substrates. Dissolved oxygen averaged 6.9 mg L^{-1} and percent saturation was 81% (ANSP, 1974) (Table 2.3).

Near-surface dissolved oxygen concentrations throughout the middle Guadalupe River typically varied seasonally with temperature. Concentrations from late autumn through late winter typically ranged from 8 to 10 mg L^{-1}, while summer concentrations usually ranged between 4 and 6 mg L^{-1} (Young et al., 1972). The major exceptions occurred at Lake Dunlap and Lake McQueeny, where primary production commonly elevated daytime concentrations to more than 10 mg L^{-1}. These shallow impoundments also exhibited extremely low benthic concentrations as well as considerable diel changes in surface concentrations. In contrast, there was little diel change in dissolved oxygen at the riverine stations, including one near Seguin (station 7) (Hannan et al., 1973).

ECOSYSTEM DYNAMICS

Detritus

Little information concerning the detrital dynamics of the middle Guadalupe River was found in the literature. Riparian forests are well developed, but the importance of direct litterfall in this 45-m-wide stream is probably modest. On the other hand, overbank flooding occurred during the earlier surveys (ANSP, 1949), so periodic floodplain inundation may provide a mechanism for the transfer of floodplain production to the river. Surface runoff from agriculture may also be an important source of allochthonous organic matter.

Downstream transport of suspended and dissolved organic matter may also be important. Canyon Reservoir probably serves as an efficient organic matter trap, which would limit inputs from the upper Guadalupe. The smaller surface-release dams of the middle Guadalupe might serve as organic matter sinks or as sources. These lakes support high standing crops of macrophytes. Sloughing and senescence may provide an important source of high-quality detritus. However, the low dissolved oxygen concentrations recorded by Young et al. (1972) from the benthos of Lake McQueeny suggest that these lakes serve primarily as organic matter sinks.

Primary Production

Hard-water conditions and moderate enrichment contributed to relatively high benthic algal production in the middle Guadalupe River near Seguin. However, these benthic assemblages were apparently either limited or otherwise influenced by high turbidity and sedimentation. A filamentous green alga, *Spirogyra* sp., and several blue-green algae, including *Phormidum uncinatum* and *P. valderianum*, were abundant in shallow mud and macrophyte beds during the earlier surveys. *Cymbella* spp., *Anomoeneis exilis*, *Navicula pumila*, and *Nitzschia palea* were abundant diatoms. Another filamentous green alga, *Chara gymnopus*, was present during 1949, became abundant in 1950, but was absent during 1952, presumably because of siltation. Epilithic assemblages were poorly developed despite the availability of rocky substrates (ANSP, 1949, 1951, 1953).

Algae were abundant during the 1962 and 1966 surveys near Seguin. *Spirogyra* sp. was abundant in backwater habitats during 1962 but was considerably less common during 1966. Conversely, a variety of blue-green algae were common in muddy shallows, on macrophytes, and on woody debris during 1966, but were generally uncommon during 1962. Diatoms were common to abundant during both surveys.

Cymbella affinis, Gomphonema affine, Nitzschia amphibia, and *N. palea* were dominant in 1962, while *Amphora cymbelloides, Amphipleura pellucida, Gyrosigma* sp., *Navicula tripunctata* var. *schizonemiodes, Navicula* sp., *Nitzschia clausii, N. obtusa* var. *scalpelliformis,* and *N. paleaformis* were common to abundant during 1966 (ANSP, 1963, 1967). The 1966 survey also recorded abundant growth of a yellow-green alga, *Vaucheria* sp., mixed with two blue-green algae (*Oscillatoria splendida* and *Microcoleus lyngbyaceus*) which were extensive in the shallows as well as on woody debris and on macrophytes (ANSP, 1967).

Algal production was moderately high during the 1973 survey, but the assemblage differed considerably from those recorded previously. *Synedra rumpens,* which was not recorded in the earlier collections, was the most common diatom. Three other diatoms, *Cocconeis pediculus, Amphora ovalis* var. *pediculus,* and *Navicula* sp., were locally abundant. *Cladophora glomerata* was the most abundant green alga, and another green alga, *Rhizoclonium hieroglyphicum,* was common in woody debris. Although it was present, *Spirogyra* sp. was uncommon. With the exception of *Oscillatoria princeps,* which was moderately common in fast water, blue-green algae were also rare (ANSP, 1974).

In summary, the algal assemblages recorded during the 1949, 1950, 1952, 1962, and 1966 surveys are characterized by an abundance of *Spirogyra* and blue-green algae. In contrast, the 1973 survey was characterized by an abundance of Cladophoracea and a paucity of blue-greens and *Spirogyra.* This transition is attributable to the hurricane-associated scouring of muddy habitats during 1972 (ANSP, 1974).

Changes following these floods were also evident for aquatic macrophytes. An abundance of macrophytes, including water lilies (probably *Numphar* sp.), sweetflag (*Acorus calamus*), and sedges (Cyperaceae), were present in shallow muddy habitats prior to 1972 (ANSP, 1949, 1951, 1953, 1963, 1967) but these were apparently much reduced during 1973 (ANSP, 1974). Dense macrophyte growths were reported for several of the small hydroelectric dams within the middle Guadalupe. Water lily (*Numphar* sp.) was abundant in Lake Dunlap, Lake McQueeny, Meadow Lake, and Wood Lake, while water hyacinth (*Eichhornia crassipes*) commonly achieved nuisance status in Lake Dunlap, Meadow Lake, and Lake Gonzales. In contrast, macrophytes were typically uncommon or absent in riverine reaches of the middle Guadalupe (Hannan et al., 1973; Young et al., 1972).

Plankton collections during August 1949 at the Academy's survey site near Seguin indicated that phytoplankton were relatively uncommon and that most of the individuals collected were entrained benthic algae (ANSP, 1949). Chlorophyl *a* samples at another site near Seguin (station 7) also indicated that summer phytoplankton populations were low. Moreover, chlorophyl *a* concentrations throughout most of the middle Guadalupe River were relatively low (>0.4 mg L^{-1}) for most of the year. There were, however, three major exceptions. One was the development of lentic phytoplankton in Lake Dunlap and Lake McQueeny (upstream from Seguin). Chlorophyll *a* concentrations in excess of 0.08 mg L^{-1} were recorded in Lake McQueeny in spring and summer and in Lake Dunlap during summer (Hannan et al., 1972). The second exception was the recording of chlorophyll *a* concentrations ranging between

0.06 and 0.153 mg L^{-1} at four lotic and two lentic sites extending approximately 80 km below a sewage treatment outfall at Seguin. The third exception, with concentrations ranging between 0.04 and 0.06 mg L^{-1}, occurred during April in the reach between Lake McQueeny and Seguin. This last increase in phytoplankton biomass was associated with an April bloom in Lake McQueeny (Hannan et al., 1973; Young et al., 1972). Significantly, diel dissolved oxygen changes were relatively minor for lotic sites, which suggests that even the high planktonic biomass recorded throughout much of the river during March was accompanied by relatively little photosynthesis. In contrast, diel changes in dissolved oxygen at Lake McQueeny exceeded 2 mg L^{-1} for much of the spring and summer (Young et al., 1972).

An investigation of nutrient budgets in three middle Guadalupe lakes (Lake Dunlap, Lake McQueeny, and Lake Gonzales) indicated that on an annual basis, they served as nitrogen sinks. In general, nitrogen retention was a function of autotrophic uptake. On the other hand, low autotrophic demand and anthropogenic enrichment resulted in a net export of phosphorus. This, combined with N/P ratios that ranged from 10: 1 to 30: 1, suggested that lentic phytoplankton within the middle Guadalupe were nitrogen limited (Hannan et al., 1972). Nutrient enrichment experiments by Stanley et al. (1990) in three shallow surface-release impoundments in the upper Guadalupe also indicated that lentic phytoplankton were largely nitrogen limited. In contrast, lotic periphyton from the upper Guadalupe were phosphorus limited.

Macroinvertebrate Production

The prevalence of mud in a variety of benthic habitats at the Academy's survey site near Seguin has had a substantial impact on the macroinvertebrate fauna. From 1949 through 1966, qualitative collections indicated that benthic habitats were characterized by an abundance of burrowing mayflies (primarily *Hexagenia bilineata*) and freshwater mussels (Unionidae) as well as by the presence of oligochaetes, an amphipod (*Hyalella azteca*), burrowing dragonflies (Gomphidae and Macromiidae), silt-tolerant mayflies (e.g., *Caenis* and *Tricorythodes*), and chironomids. Macrophyte beds supported several gastropods (e.g., *Pseudosuccinea columella*), damselflies (e.g., *Argia* and *Enallagma*), and numerous chironomids (ANSP, 1949, 1951, 1953, 1963, 1967). Dredge samples conducted during 1962 yielded *Hexagenia*, chironomids, and a freshwater mussel, *Anodonta* (ANSP, 1963).

The scouring of mud by hurricane stormflows during 1972 resulted in dramatic changes by the time the survey site was sampled in 1973. Taxonomic richness increased substantially for Ephemeroptera, Coleoptera, and Trichoptera. Densities also increased for several taxa. Damselflies (Zygoptera), Ephemeroptera, and Trichoptera were abundant in all aquatic habitats except muddy backwaters, while chironomids were abundant on woody debris. *Hydropsyche orris* and *Nectopsyche* sp. were the two most abundant trichopterans. The mayflies experienced a shift in abundant taxa. Formerly, the fauna was characterized by *Hexagenia, Caenis*, and *Tricorythodes*. These taxa were also recorded during this survey, but the 1973 fauna was best characterized by other taxa, including *Baetis* spp., *Stenonema* spp., *Isonychia* sp., and *Ameletus ludens. Corydalus cornutus*, which is characteristic of rocky streams, was recorded for the first time during 1973. *Hyalella azteca*, which was common in earlier surveys, was also common in 1973 (ANSP, 1974).

Another significant change in the pre-1972 and 1973 surveys concerned bivalves. The earlier surveys reported a diverse and abundant assemblage of unionids (ANSP, 1949, 1951, 1953, 1963, 1967). This assemblage experienced a major decline in abundance and a less serious decline in taxonomic richness by the time the site was surveyed in 1973. This decline was apparently the result of intensive scouring. Two hardy species *Amblema perplicata* and *Anodonta grandis*, were still actively reproducing. Confounding the flood-associated decline in abundance and diversity was the recording of the introduced asiatic clam (*Corbicula fluminea*). Elsewhere, native unionids have suffered significantly because of this species' superior reproductive and competitive capabilities (ANSP, 1974).

Fish Production
A total of 10, 14, and 13 species of fish were collected at the Academy's survey site in the Guadalupe River near Seguin during 1949, 1950, and 1952, respectively (ANSP, 1949, 1951, 1953). No comments of relative abundance were presented for the first two surveys, but gizzard shad (*Dorosoma cepedianum*), red shiner (*Notropis lutrensis*), and Rio Grande perch (*Cichlasoma cyanoguttatum*) were the three most abundant taxa (ANSP, 1953). A total of 10 species were collected in both the 1962 and 1973 surveys. *C. cyanoguttatum* was dominant in 1962 (ANSP, 1963) but was apparently much less common in 1973. Instead, *N. lutrensis*, bluegill (*Lepomis macrochirus*), and mosquitofish (*Gambusia affinis*) were the most common fishes (ANSP, 1974).

A total of 31 species belonging to 10 families were collected in the Guadalupe near Seguin. As would be expected in a southern warm-water stream, Cyprinidae (11 species) and Centrarchidae (10 species) accounted for most of the taxa. However, three families that one would have expected to be characteristic of this reach of the Guadalupe, Lepisosteidae (gas), Catostomidae (suckers and redhorses), and Ictaluridae (catfish), were either poorly represented or absent. Lepisosteidae were never collected at Seguin, although local residents reported that gars were present (ANSP, 1951). Catostomidae was represented by only one species, gray redhorse (*Moxostoma congestum*), which was collected only during the 1950 and 1962 surveys (ANSP, 1951, 1963). One ictalurid, channel catfish (*Ictalurus punctatus*), was collected infrequently during the 1973 survey. However, observations of the catches of local fishermen indicated that *I. punctatus* was abundant in the deeper waters, which were not sampled effectively during the surveys (ANSP, 1974).

The paucity or absence of some fishes in the Guadalupe near Seguin was attributed to the barriers to upstream movement created by the series of small hydroelectric dams (ANSP, 1949, 1963). The potential of these dams as barriers is supported by the findings that several species from these three families, including spotted gar (*Lepisosteus oculatus*), longnose gar (*L. osseus*), river carpsucker (*Carpiodes carpio*), *Moxostoma congestum*, *Ictalurus punctatus*, and flathead catfish (*Pylodictis olivaris*), were collected with regularity in the lower Guadalupe near Victoria (ANSP, 1949, 1951, 1953, 1963, 1974). However, these same species were also regularly collected in the upper Guadalupe (upstream from the Comal River). In particular, *L. osseus*, *C. carpio*, *M. congestum*, *I. punctatus*, and *P. olivaris* were widely distributed and apparently abundant in the upper river. Other species which were also frequently collected near Seguin, including *Dorosoma cepedianum*, *Notropis lutrensis*, bullhead minnow (*Pimephales vigilax*), *Gambusia affinis*, warmouth (*Lepomis gulosus*), *L. macrochirus*, longear sunfish (*L. megalotis*), and southwestern logperch (*Percina caprodes carbonaria*), were also common or abundant and widely distributed in the upper Guadalupe (Hubbs et al., 1953).

The fish fauna of the Guadalupe and adjacent drainages (i.e., San Antonio, Colorado, and the Little River subdrainage of the Trinity) is strongly influenced by the cooler, permanent spring-influenced waters of the Edwards Plateau and Balcones Fault. Guadalupe bass (*Micropterus treculi*), which was formerly considered a subspecies of spotted bass (*M. punctulatus*), is endemic to these drainages. Two other important species, Texas shiner (*Notropis amabilis*) and *Moxostoma congestum*, are most extensively distributed within these drainages, but they do occur in other Texas drainages (Hubbs, 1957; Lee et al, 1980). Finally, several endemic species of *Gambusia* are or had been restricted to major springs within the Edwards Plateau and/or the large springs along the Balcones Fault (Hubbs et al., 1953; Lee et al., 1980).

The significant influence of springs in the Guadalupe River has also facilitated the establishment of several introduced species. Rock bass (*Ambloplites rupestris*) was first stocked in the Guadalupe during 1987. Since then, this characteristically northern fish has become well established in cool, clear habitats throughout the drainage (e.g., Comal River) (Brown, 1953; Hubbs et al., 1953). The establishment of Texas populations of an eastern species, redbreast sunfish (*Lepomis auritus*), is limited primarily to those drainages influenced by the Edwards Plateau (Hubbs, 1957). The moderate-temperature regimes of the large springs have also facilitated the establishment of another introduced species, *Cichlasoma cyanoguttatum*. This cichlid is most abundant in the vicinity of the large springs where winter temperatures are tolerable (Hubbs et al., 1953), and its population increases considerably following mild winters (Brown, 1953). Two other introduced cichlids, Mozambique tilapia (*Tilapia mossambica*) and blue tilapia (*T. aurea*), are also established in the Guadalupe River (Lee et al., 1980). Sailfin molly (*Poecilia latipinna*), which is abundant in the Guadalupe below the Comal River, was once considered to be an introduced species (Brown, 1953), but Lee et al. (1980) reported that is probably native.

COMMUNITIES OF AQUATIC LIFE

These studies of riverine systems support the theory of Eugene P. and Howard T. Odum that in each stable community there are four functional stages of nutrient and energy transfer: (1) detritus producers and primary producers, (2) detritivores and herbivores; (3) omnivores, and (4) carnivores.

Functional Relationships

The relative importance of allochthonous and autochthonous inputs into the middle Guadalupe River are not known. The combination of extensive siltation, turbidity, and limited littoral habitat probably substantially reduced the importance of benthic algae as a source of tropic support. However, attached algae from muddy backwaters, macrophyte beds, and woody debris probably provided food for localized assemblages of primary consumers, including some gastropods and chironomids. The increase in the abundance of algal-grazing mayflies (e.g., *Ameletus* and *Stenonema*) and the herbivorus shredder *Nectopsyche* during the 1973 survey suggests that benthic algal production became important following scouring hurricane storm flows.

The conditions that prevailed in the middle Guadalupe prior to the 1972 storm flows resulted in considerable deposition of fine sediments. Presumably, these depositional conditions also resulted in the accumulation of fine detritus, which would have provided a major source of tropic support for benthic macroinvertebrates. This is suggested by the abundance of several benthic detritivores, including *Hexagenia*, *Caenis*, *Tricorythodes*, chironomids, and oligochaetes.

Organic matter within the water column (i.e., organic seston) is another important source of trophic support in the middle Guadalupe. The source of this organic seston could include the export of lentic phytoplankton and zooplankton, the export of detritus from lentic macrophytes, allochthonous inputs from inundated floodplains, or agricultural runoff. Unionids were the primary macroinvertebrate filter feeders in the middle Guadalupe prior to the 1972 storm flows. They would have filtered smaller particulates, including fine detritus and phytoplankton. Unionids were reduced following the storm flows, and a net-spinning caddisfly, *Hydropsyche orris,* became the dominant filter feeder. A filter-feeding mayfly, *Isonychia,* a filter-feeding dipteran, *Simulium* near *venustum,* and the introduced clam, *Corbicula fluminea,* were also recorded at this time. The remaining mussels, *Corbicula, Isonychia,* and *Simulium,* would have fed on fine detritus and/or phytoplankton. The abundant *H. orris,* on the other hand, would probably have selected a more varied diet, which would have included larger pieces of detritus, zooplankton, drifting macroinvertebrates, as well as fine detritus and phytoplankton.

Damselflies, including *Argia* and *Enallagma*, appeared to be the most abundant macroinvertebrate predators. Burrowing dragonflies, including *Gomphus* and *Macromia*, also appeared to be abundant. Other, apparently less abundant, predators included leeches (Hirudinoidia), assorted Hemiptera, *Corydalus cornutus*, *Sialis* (Megaloptera), *Oecetis* (Trichoptera), and Tanypodinae (Chironomidae).

The fish of the middle Guadalupe contained a number of omnivorous bottom feeders, which could feed on benthic detritus, algae, macrophytes, and/or invertebrates. One of the more abundant of these was *Notropis lutrensis*. The introduced cichlid, *Cichlasoma cyanoguttatum*, has exhibited considerable plasticity in its diet. It has been reported to be exclusively herbivorous in the heavily vegetated outlet from San Marcos Springs, omnivorous in the Rio Grande, and primarily detritivorous in northeastern Mexico. One possibility is that the diet shifts of this opportunistic species are influenced by competition from ecologically equivalent centrarchids. *Carpiodes carpio* is another bottom-feeding omnivore. The diet of *Moxostoma congestum* has not been studied, but several other species of *Moxostoma* are bottom-feeding omnivores (Lee et al., 1980; Pflieger, 1975).

Dorosoma cepedianum undergoes ontogenic shifts in its diet, with phytoplankton and zooplankton being important diet components in younger fishes and benthic invertebrates dominating the diets of older fishes. Terrestrial drift, aquatic insects, entomostracans, and algae are eaten by *Gambusia affinis, Poecilia latipinna,* and *Fundulus notatus* (Lee et al., 1980; Pflieger, 1975).

Pimephales vigilax and *Percina caprodes carbonaria* feed primarily on benthic invertebrates. Others, including *Lepomis macrochirus, L. gulosus,* and *L. megalotis,* feed heavily on invertebrates but also feed on small fishes. *Lepisosteus* spp. and *Pylodictis olivaris* shift from a diet dominated by invertebrates while young to one dominated by fish when older. A similar pattern is exhibited by *Ictalurus punctatus,* although it will also feed on carrion and plant material. The diet of *Micropterus treculi* has apparently not been studied, but that of its closest relative, *M. punctulatus,* consists primarily of crayfish, insects, and fish (Lee et al., 1980; Pflieger, 1975).

Structure of Aquatic Ecosystems

The aquatic communities near Seguin have been studied by the Academy of Natural Sciences since 1949. The years of study are given at the end of the table of species (Table 2.4). These were the only studies that were complete so that the structure of the community could be constructed. To illustrate the structure of the community, the 1949 study was chosen because at that time there were very few dams on the Guadalupe River and there were no evident sources of pollution entering the river above this area. Downstream there were many instances where salt water from oil wells entered the river, but these did not influence as such the community where the study was made.

Vegetative Habitats. In 1949 there was a well-developed algal flora in the river. In and among the vegetation were the blue-green algae *Lyngbya aestuarii* and *Oscillatoria splendida*. Several green algae were found in and among the vegetation. Among the vegetation, which consisted of unidentified species of the Cyperaceae and *Acorus calamus,* were *Oedogonium* sp., *Spirogyra* spp., and *Vaucheria* sp. Also present with emergent leaves was an unidentified species of the family Nymphaceae.

Several diatoms were found attached to the vegetation or in debris associated with the vegetation: *Synedra acus, S. ulna, S. ulna* var. *danica, Achnanthes affinis, A. exigua, Cymbella microcephala, C. ventricosa, Cymbella* spp., *Gomphonema intricatum* var. *vibrio, G. lanceolata, G. parvulum, Anomoeoneis exilis,* and *Anomoeoneis* sp. Also growing on the vegetation were *Gyrosigma spencerii* var. *nodifera, Navicula cryptocephala* var. *pumila, N. gregaria, N. grimmei, N. lanceolata, N. radiosa, N. radiosa* var. *tenella,* and some unidentified species of *Navicula*. Various Nitzschias were found in the debris associated with the vegetation: *Nitzschia amphibia, N. frustulum* var. *tenella,* and *N. palea*.

Several protozoans were found associated with the vegetation. They were the primary producer, *Cryptomonas erosa*; *Peranema trichophorum*; and detritivores that feed mainly on bacteria, *Bodo alexeieffii, B. caudatus,* and *Bodo* spp. Crawling over

(text continues on page 96)

TABLE 2.4.   *Species List: Guadalupe River*

Taxon[a]	Substrate[b] V	Lentic M	Lentic S	Lotic G	Lotic R	Source[c]
SUPERKINGDOM PROKARYOTAE						
KINGDOM MONERA						
Division Cyanophycota						
Class Cyanophyceae						
Order Chroococcales						
Family Chroococcaceae						
Coccochloris elabens						c
Family Entophysalidaceae						
Entophysalis rivularis		X				d, f
Order Nostocales						
Family Nostocaceae						
Cylindrospermum majus		X				b
* *Cylindrospermum* sp.		X				a
Family Oscillatoriaceae						
Lyngbya aerugineo-caerulea	X				X	c
* *L. aestuarii*	X	X				a,e
L. putealis	X				X	d
Microcoleus chthonoplastes	X					b–d
M. lyngbyaceus	X	X				e
Oscillatoria chalybea		X				d,e
O. limosa	X	X				c
O. lutea	X				X	f
* *O. princeps*		X			X	a–f
O. proboscidea		X				c
* *O. spendida*	X	X	X	X	X	a,b
* *O. tenuis*		X			X	a,b,e
Phormidium unicatum		X			X	c
P. valderianum		X				c
Schizothrix arenaria		X	X			e,f
S. calcicola		X	X			d–f
S. friesii		X				f
S. mexicana					X	e
S. rubella		X				e
Symploca muscorum		X				e
Family Rivulariaceae						
Calothrix parietina		X				d
SUPERKINGDOM EUKARYOTAE						
KINGDOM PLANTAE						
SUBKINGDOM THALLOBIONTA						
Division Chlorophycota						
Class Chlorophyceae						
Order Tetrasporales						
Family Tetrasporaceae						
Tetraspora gelatinosa	X	X			X	f
Order Ulotrichales						

TABLE 2.4. (*Continued*)

Taxon[a]	Substrate[b]					Source[c]
	Lentic			Lotic		
	V	M	S	G	R	
Family Ulothrichaceae						
Ulothrix zonata	X					c
Order Oedogoniales						
Family Oedogoniaceae						
* *Oedogonium* sp.	X	X				a,b,d,f
Order Siphonocladales						
Family Cladophoraceae						
* *Cladophora glomerata*					X	a,f
* *Cladophora* sp.					X	a,c
Pithora sp.					X	d
* *Rhizoclonium hieroglyphicum*					X	b,d,f
Order Siphonales						
Family Dichotomosiphonaceae						
Dichotomosiphon tuberosus	X				X	c
Order Zygnematales						
Family Desmidiaceae						
Cosmarium subcostatum	X	X				c
Penium margaritaceum	X	X				c
Family Zygnemataceae						
Mougeotia sp.					X	c
* *Spirogyra* spp.	X	X	X	X	X	a–e
Class Charophyceae						
Order Charales						
Family Characeae						
* *Chara gymnopus*		X	X			a,b
Chara sp.		X	X			d
Division Chromophycota						
Class Xanthophyceae						
Order Vaucheriales						
Family Vaucheriaceae						
Vaucheria sessilis	X	X	X	X	X	b
* *Vaucheria* sp.	X	X	X			a,d–f
Class Bacillariophyceae						
Order Eupodiscales						
Family Coscinodiscaceae						
Cyclotella meneghiniana	X	X	X			d,f
C. striata	X	X	X			d
Melosira varians	X			X	X	f
Order Biddulphiales						
Family Biddulphiaceae						
Biddulphia laevis	X	X				f
Order Fragilariales						
Family Fragilariaceae						
Diatoma vulgaris	X					f
Fragilaria pinatta	X					f

(*continued*)

TABLE 2.4.   (*Continued*)

Taxon[a]	Substrate[b]					Source[c]
	Lentic			Lotic		
	V	M	S	G	R	
Family Fragilariaceae (*cont.*)						
F. vaucheria	X	X				c,f
Opephora sp.			X			c
* *Synedra acus*	X	X	X	X		a
S. affinis v. *obtusa*	X	X				f
S. goulardi	X	X				c,f
S. minuscula	X					f
S. rumpens	X					f
* *S. ulna*	X			X		a,b,d,f
* *S. ulna* v. *danica*	X			X		a,c
S. ulna v. *oxyrhynchus*	X	X	X		X	f
Order Achnanthales						
Family Achnanthaceae						
* *Achnanthes affinis*	X	X	X		X	a,b
* *A. exigua*	X	X	X			a
A. minutissima	X					c,d,f
A. minutissima v. *cryptocephala*	X					b
Cocconeis pediculus	X				X	f
C. placentula v. *euglypta*	X				X	d,f
Rhoicosphenia curvata	X				X	f
Order Naviculales						
Family Cymbellaceae						
Amphora cymbelloides	X				X	e
A. ovalis					X	d
A. ovalis v. *pediculus*		X	X			d,f
A. pediculus	X					c
A. perpusilla	X					c
Amphora sp.		X	X			d
* *Cymbella affinis*		X	X			a–f
* *C. cessatii*		X	X			a
* *C. hybridiformis*		X	X			a
* *C. microcephala*	X					a–d
C. pusilla		X	X	X		d
C. sinuata	X					f
C. tumida	X	X	X			c,f
C. turgida	X			X		d,f
C. turgidula				X	X	c
* *C. ventricosa*	X		X			a,b,d,f
* *Cymbella* spp.	X			X		a–c
Family Gomphonemaceae						
Gomphonema affine	X					d,f
G. intricatum	X	X				f
G. intricatum v. *pumila*	X	X				f
* *G. intricatum* v. *vibrio*	X	X				a
* *G. lanceolata*	X				X	a

TABLE 2.4. (*Continued*)

Taxon[a]	V	Lentic		Lotic		Source[c]
		M	S	G	R	
Family Gomphonemaceae (*cont.*)						
* G. parvulum	X	X	X		X	a,c,d,f
G. sphaerophorum	X				X	d
Gomphonema spp.					X	b,c
Family Naviculaceae						
Amphipleura pellucida	X					b–e
* Anomoeoneis exilis	X					a–d
A. exilis v. thermalis	X					b
A. pellucida						a
* Anomoeoneis sp.	X	·				a
Caloneis bacillum	X					d
C. hyalina	X					f
C. schumanniana	X					d
C. silicula v. truncatula	X					d
Caloneis sp.	X	X	X			b,c
Diploneis pseudovalis	X					d
D. puella	X					c,d
Gyrosigma attenuatum		X	X			d
* G. kuetzingii		X	X			a,c,d
G. nodiferum		X	X			f
G. scalproides	X					d
G. spenceri v. curvula	X					b
* G. spenceri v. nodifera	X	X	X			a–c
Gyrosigma sp.	X					e
Navicula canalis	X	X				d
N. cincta	X	X				d,f
N. confervaceae	X	X				d
* N. cryptocephala v. pumila	X	X	X			a
N. cryptocephala v. veneta	X	X				d
N. cuspidata v. ambigua	X	X				b,d
N. germainii	X	X				d
N. gracilis v. schizonemoides	X					d
* N. gregaria	X	X	X			a
* N. grimmei	X	X	X			a
N. halophila		X	X			d
* N. kotachyi		X	X			a
* N. lanceolata	X			X		a,c,d
N. luzonensis	X			X		f
N. minima		X	X			d
N. mutica		X	X			c
N. pumila		X	X			c
* N. pupula		X	X			a
N. pupula v. capitata	X			X		b,d
N. pupula v. rectangularis		X	X			d
* N. radiosa	X					a

(*continued*)

TABLE 2.4. (*Continued*)

Taxon[a]	Substrate[b]					Source[c]
		Lentic		Lotic		
	V	M	S	G	R	
Family Naviculaceae (*cont.*)						
* *N. radiosa* v. *tenella*	X	X	X[e]			a
N. rhynchocephala v. *germainii*	X	X				f
N. ruttneri v. *rostrata*	X	X				d
N. salinarum v. *intermedia*	X	X	X	X		f
N. savannahiana	X	X				d,f
N. seminulum	X	X				d,f
N. subatomoides	X	X				c
N. symmetrica	X	X				d,f
N. tenella	X	X				b
N. thienemanni	X	X				c
N. tripunctata v. *schizonemoides*	X	X				e
* *Navicula* spp.	X	X				a–f
Pinnularia gibba v. *subundulata*						d
P. microstauron	X	X				c
Order Bacillariales						
Family Nitzschiaceae						
Bacillaria paradoxa	X	X	X			c
Nitzschia acicularis				X	X	d
* *N. amphibia*	X	X	X			a,b,d,f
N. bacata	X	X				d
N. clausii	X	X				c,e
N. confinis		X				d,f
N. dissipata		X				f
N. filiformis				X		c,f
N. frustulum				X		c,d
* *N. frustulum* v. *minutula*				X		b
N. frustulum v. *perminuta*				X		f
N. frustulum v. *subsalina*				X		d,f
* *N. frustulum* v. *tenella*	X			X		a
N. gracile				X	X	d
N. hungarica		X				d
N. kutzingiana		X				c,d
N. obtusa v. *scalpelliformis*	X	X		X	X	d,e
* *N. palea*	X	X		X	X	a–f
N. paleacea	X	X				b
N. paleaformis	X	X				e
N. sigma	X	X				d
N. sigma v. *rigida*	X	X				c
N. subvitrea	X					c
N. triblionella v. *debilis*		X				d
N. triblionella v. *levidensis*		X				d
N. triblionella v. *victoriae*		X				f
Order Surirellales						

TABLE 2.4. (*Continued*)

Taxon[a]	V	Lentic		Lotic		Source[c]
		M	S	G	R	
Family Surirellaceae						
Surirella robusta		X				d
Class Dinophyceae						
Order Peridinales						
Family Cystodiniidae						
Glenodinium neglectum				X	X	d
Family Peridiniaceae						
Peridinium sp.				X	X	c
Division Rhodophycota						
Class Rhodophyceae						
Order Compsopogonales						
Family Compsopogonaceae						
Compsopogon coeruleus					X	d
Order Acrochaetinales						
Family Acrochaetiaceae						
* *Chantransia* sp.					X	a
Order Nemalionales						
Family Batrachospermaceae						
Batrachospermum sp.				X	X	e,f
SUBKINGDOM EMBRYOBIONTA						
Division Magnoliophyta						
Class Dicotyledoneae						
Order Nymphaeales						
Family Numpheaeceae						
* unident. sp.	X					a–c
Class Monocotyledoneae						
Order Arales						
Family Araceae						
* *Acorus calamus*	X	X				a–d
Order Cyperales						
Family Cyperaceae						
* unident. spp.	X	X				a–d
KINGDOM ANIMALIA						
SUBKINGDOM PROTOZOA						
Class Mastigophora						
Order Chrysomonadida						
Family Chromulinidae						
Chromulina sp.		X	X	X	X	f
Family Ochromonadidae						
Anthophysa steinii	X					c
Anthophysis vegetans	X					d,f
Monas guttula		X	X			d
Order Cryptomonadida						
Family Cryptomonadidae						
o *Chilomonas paramecium*		X	X			c

(*continued*)

TABLE 2.4.   (*Continued*)

Taxon[a]	V	Lentic M	Lentic S	Lotic G	Lotic R	Source[c]
Family Cryptomonadidae (*cont.*)						
Chroomonas sp.		X	X			f
Cryptochrysis commutata	X					f
*o *Cryptomonas erosa*	X					a,c,d,f
o *C. marssonii*		X	X			c
o *C. ovata*		X	X			f
Cyathomonas truncata		X				f
Order Phytomonadida						
Family Carteriidae						
Spermatozopsis exultans		X				f
Family Chlamydomonadidae						
Chlamydomonas globosa	X					d,f
C. gracilis	X					d
Chlamydomonas sp.	X					c
Scourfieldia complanata	X					d
Family Phacotidae						
Phacotus lenticularis		X	X			c,d,f
Family Volvocidae						
Eudorina elegans		X	X			d,f
Gonium pectorale		X	X			d
G. sociale		X	X			d
Pandorina morum		X	X			d,f
Order Euglenoidida						
Family Anisonemidae						
Anisonema acinus	X	X	X			c,d,f
Entosiphon sulcatum		X	X			c,d
Heteronema spirale		X	X			d
H. tremulum		X	X			d
Notosolenus apocamptus	X					d
Peranema granulifera	X					f
*o *P. trichophorum*	X	X				a,c,d,f
Family Astasiidae						
*o *Astasia klebsii*		X				a,d
Petalomonas mediocanellata	X					d,f
Scytomonas pusilla						d
Family Euglenidae						
Ascoglena vaginicola		X	X			f
Cryptoglena pigra		X	X			d
Euglena agilis		X	X			c
E. chadefaudii		X	X			c
E. deses	X					d
E. gracilis	X					d,f
E. haematodes		X	X			d
E. intermedia		X	X			c
E. minima	X					d
E. mutabilis		X	X			c

TABLE 2.4.   (*Continued*)

Taxon[a]	V	Lentic M	S	Lotic G	R	Source[c]
Family Euglenidae (*cont.*)						
E. oxyuris	X					d
E. oxyuris v. *charkowiensis*	X					c
* *E. pisciformis*		X	X			a,f
E. rubra		X				f
* *E. viridis*		X	X			a–d
Lepocinclis texta		X				c,f
Phacus pleuronectes		X				c,d
P. pyrum		X				c,d,f
P. triqueter		X				f
Trachelomonas hispida		X				d
T. schauinslandii		X				c
T. vapiformis v. *elegans*		X				c
T. volvocina		X				d
Order Rhizomastigida						
Family Mastigamoebidae						
Mastigamoeba auriculata						d
Order Protomonadida						
Family Amphinomadidae						
Cladomonas fruticulosa						c
Family Bodonidae						
*do *Bodo alexeieffii*	X	X	X			a
B. amoebinus	X	X	X			d
*do *B. caudatus*	X	X	X			a,d
B. fusiformis	X	X				c
B. obovatus	X	X				d
B. putrinus	X	X				f
B. saltans	X	X				d
*do *Bodo* spp.	X	X				a
Pleuromonas jaculans	X	X	X			f
Class Sarcodina						
Order Proteomyxida						
Family Vamprellidae						
Nuclearia simplex	X					f
Order Amoebida						
Family Amoebidae						
Acanthamoeba hylina	X	X				f
do *Amoeba dubia*	X	X				f
do *A. gorgonia*	X	X				f
do *A. limnicola*	X	X				d
*do *A. proteus*	X	X				a,d
*do *A. radiosa*	X	X				a
do *A. spumosa*	X	X				d,f
do *A. striata*	X	X				c,f
*do *A. verrucosa*	X					a,f

(*continued*)

TABLE 2.4. (*Continued*)

Taxon[a]	V	M	S	G	R	Source[c]
Family Amoebidae (*cont.*)						
do　*A. vespertilia*	X	X				f
do　*Amoeba* sp.	X	X				c
*d　*Vahlkampfia limax*				X		a,d,f
Order Testacida						
Family Difflugiidae						
do　*Difflugia acuminata*		X				c
*do　*D. globulus*	X					a
*do　*D. lobostoma*	X	X				d
do　*D. oblonga*	X					c,d,f
do　*Difflugia* sp.	X	X				c
Family Euglyphidae						
Cyphoderia ampulla	X					f
Family Gromiidae						
Lecythium hyalinum	X					d
Order Heliozoida						
Family Acanthocystidae						
Acanthocystis aculeata	X	X	X			f
Family Actinophryidae						
*　*Actinophrys sol*		X				a,d
Class Ciliata						
Order Gymnostomatida						
Family Amphileptidae						
Amphileptus claparedei	X					d
Lionotus fasciola		X				d,f
*do　*L. trichocystis*	X					a
Family Chlamydodontidae						
Chilodonella caudata	X	X	X			f
C. cucullulus	X	X	X			f
C. fluviatilis	X	X	X			d
C. uncinata	X	X	X			c,f
Family Colepidae						
Coleps hirtus						d,f
Coleps sp.						c
Family Didiniidae						
Didinium balbianii		X	X			d
Mesodinium acurus						f
Family Dysteriidae						
Trochilia minuta						f
Family Holophryidae						
Holophrya simplex		X	X			b,d
Lacrymaria olar						d,f
Lacrymaria sp.						d
Prorodon discolor		X	X			d
*　*Prorodon* sp.		X				a
Pseudoprorodon farctus						d
Urotricha farcta						d,f

TABLE 2.4. (*Continued*)

Taxon[a]	V	M	S	G	R	Source[c]
		Lentic		Lotic		
Family Loxodidae						
Loxodes magnus	X	X				c
L. rostrum	X	X				f
L. vorax	X	X				d
Family Nassulidae						
Chilodontopsis depressa		X	X			f
*hd *Nassula ornata*				X	X	a
Trochiloides palustris						d
T. recta						d
Family Spathiidae						
Cranotheridium taeniatum		X	X			b
Family Tracheliidae						
Dileptus anser		X	X			f
Trachelius ovum		X				b
Order Hymenostomatida						
Family Frontoniidae						
Cinetochilum margaritaceum	X	X	X			c,f
Cyrtolophosis centralis		X	X			d
Frontonia acuminata	X					c,f
F. leucas	X					c,d
Lembadion bullinum						c,f
Loxocephalus plagius	X					c
Saprophilus muscarum						d
*do *Urocentrum turbo*				X		a,d
Family Parameciidae						
* *Paramecium aurelia*		X				a,d
P. caudatum		X				c
P. multimicronucleatum		X				c
P. putrinum		X				d
P. trichium		X				d
Physalophyra spumosa		X				b
Family Pleuronematidae						
Cyclidium glaucoma		X				c,f
C. litomesum		X				d,f
Pleuronema sp.				X		d
Family Tetrahymenidae						
Glaucoma scintillans				X		d
Monochilum frontatum						f
Tetrahymena pyriformis						d
Order Heterotrichida						
Family Bursariidae						
Bursaridium difficile		X				d
Spirostomum teres						d
Order Oligotrichida						
Family Halteriidae						
Halteria grandinella					X	c,d,f

(*continued*)

TABLE 2.4.   (*Continued*)

Taxon[a]	V	M	S	G	R	Source[c]
Family Strobilidiidae						
Strobilidium gyrans					X	d,f
Order Hypotrichida						
Family Aspidiscidae						
Aspidisca costata		X				b–d
A. polystyla		X				f
Family Euplotidae						
Euplotes eurystomus		X	X			b–d,f
E. patella		X	X			d,f
Family Oxytrichidae						
Eschaneustyla brachytona	X	X	X			b
Gonostomum strenuum	X	X	X			b
Opisthrotricha similis						f
*do *Oxytricha fallax*	X	X				d
*do *O. ludibunda*	X	X				a
do *O. setigera*	X	X				d,f
do *Oxytricha* sp.				X		f
do *Stichotricha secunda*				X		f
do *Stylonychia mytilis*	X	X				b,f
do *S. pustulata*				X		f
do *S. putrina*				X		d
do *Trichotaxis fossicola*	X					c
do *Uroleptus limnetis*	X			X		d,f
do *Urosoma acuminata*	X	X				b
do *U. caudata*	X	X				d
do *U. cinkowski*	X	X				c
do *Urostyla grandis*	X	X				f
Order Peritrichida						
Family Vorticellidae						
*do *Vorticella convallaria*	X			X	X	a,d
do *V. microstoma*	X			X	X	d
do *V. montilata*	X			X	X	d
do *V. picta*	X			X	X	f
do *Vorticella* sp.	X			X	X	c
SUBKINGDOM PARAZOA						
Phylum Porifera						
Class Demospongiae						
Order Haplosclerida						
Family Spongillidae						
o *Spongilla* sp.				X	X	c
Trochospondilla horrida	X				X	d
unident. sp.						d
SUBKINGDOM EUMETAZOA						
Phylum Cnidaria						
Class Hydrozoa						
Order Hydroida						

TABLE 2.4. (*Continued*)

Taxon[a]	V	M	S	G	R	Source[c]
		Lentic		Lotic		
Family Clavidae						
Cordylophora lacustris	X				X	d
Phylum Platyhelminthes						
Class Turbellaria						
Order Tricladida						
Family Planariidae						
do *Dugesia tigrina*	X	X			X	b–d,f
Phylum Nemertea						
Class Enopla						
Order Hoplonemertea						
Family Tetrastemmatidae						
Prostoma rubrum	X					f
Phylum Rotifera						
Class Digonota						
Order Bdelloidea						
unident. sp.						b
Family Philidonidae						
Philodina citrina						a
P. roseola						c
Class Monogononta						
Order Ploima						
Family Brachionidae						
Brachionus uroceolaris	X	X	X			b
Keratella cochlearis	X					b
Family Colurellidae						
Lepadella patella		X	X			c
Family Euchlanidac						
*o *Euchlanis parva*	X					a
*o *Tripeuchlanis plicata*	X	X				a
Family Lecanidae						
Lecane bulla						c
L. luna						b
*o *L. luna* v. *presumpta*	X	X				a
o *L. lunaris*						c
o *L. papuana*						b,c
o *L. pyriformis*						c
*o *L. stichaea*	X	X				a
Monostyla bulla		X				b
* *M. cornuta*	X	X				a
* *M. hamata*		X				a
* *M. lunaris*		X				a
* *M. pyriformis*		X				a,b
Monostyla sp.		X				b
Family Notommatidae						
* *Cephalodella auriculata*	X	X				a
* *C. forficula*	X	X				a
* *Cephalodella* sp.	X	X				a

(*continued*)

TABLE 2.4.   (*Continued*)

Taxon[a]	V	Lentic M	Lentic S	Lotic G	Lotic R	Source[c]
Family Synchaetidae						
Polyarthra sp.						a
Family Trichoceridae						
* *Trichocerca rattus*		X	X			a
Phylum Mollusca						
Class Gastropoda						
Order Mesogastropoda						
Family Hydrobiidae						
* *Amicola limosa*	X					a
Amnicola sp.			X	X		c,d
Amnicola (Cincinatia) sp.			X	X		b
Lyrodes coronatus			X	X		b,c
Potamopygus coronatus			X	X		d,f
Order Basommatophora						
Family Ancylidae						
Ferrisia excentrica (= *L. fuscus?*)	X	X				b,c
Laevapex fuscus	X	X				f
Family Lymnaeidae						
Lymnaea humilis		X	X			c
L. humilis modicella		X	X			b
*do *Pseudosuccinea columella*	X					a–c
Family Physidae						
Physa halei	X	X	X			b,c
*do *P. integra*	X	X	X			a
Physa sp.		X	X			f
Family Planorbidae						
* *Helisoma anceps*	X					a,b,f
H. trivolvis	X					d,f
Menetus sp.				X		f
unident. spp.						f
Class Bivalvia						
Order Unionoida						
Family Unionidae						
Amblema costata (= *a. peprlicata?*)			X	X		d,e
*o *A. perplicata*			X	X		f
*hd *A. plicata perplicata* (= *A. peprlicata?*)	X		X	X		a–c
Anodonta grandis			X	X		b–f
A. imbecillis			X	X		d–f
Carunculina parva texasensis			X	X		b–e
Cyrtonaias tampecoensis			X	X		f
* *Lampsilis anodontoides*			X	X		a–e
* *L. fasciata hydiana* (*L. hydiana?*)			X	X		a–e
L. hydiana			X	X		f
L. (= *Cyrtonaias?*) *tampecoensis*			X	X		d,e
*o *Proptera pupurata* (= *Cyrtonaias tampecoensis?*)			X	X		a–c
*o *Quadrula aurea*			X	X		a–f

TABLE 2.4. (*Continued*)

Taxon[a]		V	M	S	G	R	Source[c]
			Lentic		Lotic		
	Family Unionidae (*cont.*)						
	Q. quadralupensis			X	X		b
*hd	*Q. petrina*			X	X		a,b
*hd	*Q. quadrula apiculata (= Q. quadrula?)*			X	X		a–e
	Q. quadrula			X	X		f
*hd	*Q. (= Tritogonia?) verrucosa*			X	X		a–c
	Quadrula sp.			X	X		a
	Tritogonia verrucosa			X	X		d,e
	Order Veneroida						
	Family Corbiculidae						
o	*Corbicula manilensis (= fluminea?)*				X	X	f
	Family Sphaeriidae						
*do	*Eupera cubensis*	X	X				f
*do	*E. singleyi (= cubensis?)*	X	X				d
	Pisidium sp.				X	X	f
o	*Sphaerium transversum*		X		X		f
*o	*Sphaerium* sp.		X		X		a,c,e
	Phylum Annelida						
	Class Hirudinoidia						
	Order Rhynchobdellida						
	Family Glossiphoniidae						
c	*Helobdella lineata*	X					f
c	*H. stagnalis*	X					b,d
	Placobdella sp.	X	X				b,c
	Order Arhynchobdellida						
	Family Erpobdellidae						
c	*Erpobdella punctata*	X					f
	Class Oligochaeta						
	unident. sp.						a,b
	Order Lumbriculida						
	Family Lumbriculidae						
	unident. sp.		X	X			c
	Order Haplotaxida						
	Family Megascolecidae						
	unident. sp.	X					f
	Family Tubificidae						
do	*Branchiura sowerbyi*		X	X			b,d
	unident. sp.			X			f
	Phylum Arthropoda						
	Class Arachnida						
	Order Acariformes						
	unident. spp.	X					f
	Class Crustacea						
	Subclass Ostracoda						
	unident. spp.						b
	Subclass Malacostraca						
	Order Amphipoda						

(*continued*)

TABLE 2.4.   (*Continued*)

Taxon[a]	V	M	S	G	R	Source[c]
Family Talitridae						
o *Hyalella azteca*	X	X				b,c,f
Order Decapoda						
Family Cambaridae						
c *Cambarus (= Procambarus?) clarki*	X	X	X			b,c
c *Cambarus* sp.	X					f
c *Procambarus clarki*	X	X	X			d
Family Palaemonidae						
o *Palaemonetes paludosa*	X	X	X			b,c,f
o *Palaemonetes* sp.	X	X	X			e
Family Penaeidae						
*o *Penaeus* sp.						a
Family Potamogiidae						
unident. sp.						a
Class Insecta						
Order Odonata						
Suborder Zygoptera						
Family Calopterygidae						
*c *Hetaerina americana*	X	X	X		X	f
*c *H. titia*	X	X	X		X	b,f
*c *Hetaerina* sp.	X	X	X		X	a
Family Coenagrionidae						
c *Argia apicalis*	X	X			X	b,d
c *A.* prob. *immuda*	X	X			X	c
*c *A. moesta*	X	X			X	a,b,d
c *A. sedula*	X	X			X	b,d,f
*c *A. translata*	X	X			X	a,b,d,f
*c *Argia* spp.	X	X			X	a,b,e
c *Enallagma basidens*	X	X				d
c *E. civile*	X	X				b
c *E. novachispaniae*	X	X				f
c *E. signatum*	X	X				d
c *E. vesperum*	X	X				d
c *Enallagma* sp.	X	X				e,f
c *Ischnura* sp.	X					c
*c *Neonura aaroni?*	X					a
Suborder Anisoptera						
Family Corduliidae						
c *Epicordulia princeps*	X	X				d
Family Gomphidae						
c *Dromogomphus spinosus*		X	X	X	X	b,c
c *D. spoliatus*		X	X	X	X	b,f
c *Gomphoides albrighti?*		X	X			f
c *Gomphoides* sp.		X	X			d
*c *Gomphus (Gomphurus) exilis*	X	X	X			a,b
*c *G. (Gomphurus) externus*	X	X	X			a,b

TABLE 2.4. (*Continued*)

Taxon[a]	Substrate[b]					Source[c]
	Lentic			Lotic		
	V	M	S	G	R	
Family Gomphidae (*cont.*)						
c G. (*Gomphurus*) nr. *fraterus*	X	X	X			c
c G. (*Stylurus*) *laurae*	X	X	X			f
c G. *militaris*	X	X	X			b,f
*c G. poss. *militaris*	X	X	X			a
*c G. *notatus*	X	X	X			a
c G. (*Stylurus*) *notatus?*	X	X	X			c
*c G. *olivaceous*	X	X	X			a
c *Gomphus* sp.	X	X	X			d
Family Libellulidae						
c *Celithemis* sp.	X					b
c *Dythemis velox nigrescens*				X	X	d
c *Pachydiplax longipennis*		X				b
c *Perithemis domitia*		X				b
unident. sp.						f
Family Macromiidae						
*c *Didymops transversa*		X				a
c *D*. prob. *transversa*		X				b
*c *Macromia* sp.		X				a–c,f
Order Ephemeroptera						
Suborder Schistonata						
Family Baetidae						
o *Baetis* spp.	X	X	X	X	X	f
o *Callibaetis* sp.	X					d
Family Ephemeridae						
o *Hexagenia bilineata*		X	X			a,f
o *H*. prob. *bilineata*		X	X			b,c
*o *Hexagenia* sp.		X	X			d,e
Family Heptageniidae						
o *Stenonema femoratum tripunctatum*				X		c
o *S. integrum*				X		f
o *S. interpunctatum* nr. *frontale*				X		b
o *S.* nr. *heterotarsale*				X		f
o *S.* nr. *terminatum*				X		f
o *Stenonema* sp.				X		d
Family Palingeniidae						
*o *Pentagenia vittigera?*	X	X	X			a
Family Polymitarcyidae						
o *Campsurus* sp.		X				c,d
*o *Tortopus* sp.	X	X			X	a
Family Siphlonuridae						
dh *Ameletus ludens*	X	X	X	X	X	f
co *Isonychia* sp.				X	X	f
Suborder Pannota						
Family Caenidae						
*o *Caenis* spp.			X		X	a–d

(*continued*)

TABLE 2.4.   (*Continued*)

Taxon[a]	Substrate[b]					Source[c]
		Lentic		Lotic		
	V	M	S	G	R	
Family Tricorythidae						
*o *Tricorythodes spp.*	X		X			a–d,f
Order Plecoptera						
Family Perlidae						
c *Neoperla clymene*				X	X	b
Order Hemiptera						
Family Belostomatidae						
c *Belostoma flumineum*	X	X	X			c
c *B. lunatrium*	X	X	X			b
Family Gelastocoridae						
c *Gelastocoris occulatus*	X	X	X			b
Family Gerridae						
o *Gerris gillettei*		X	X			b
o *G. marginatus*		X	X			c
o *Limnogonus hesione*	X	X	X			b,c
*o *Metrobates artus*				X	X	a,c
o *M. trux*		X	X			f
Rheumatobates hungerfordi		X	X			c
*c *Trepobates inermis*		X	X	X	X	a–d
Family Mesoveliidae						
Mesovelia bisignata	X					c
Family Nepidae						
*c *Ranatra australis*	X	X	X			a,b
R. prob. *australis*	X	X	X			d
Family Veliidae						
*c *Rhagovelia choreustes*				X	X	a,f
c *Rhagovelia* sp.				X	X	f
Order Megaloptera						
Family Corydalidae						
o *Corydalus cornutus*		X	X	X	X	f
Family Sialidae						
c *Sialis* sp.	X	X	X	X		c
Order Coleoptera						
Suborder Adephaga						
Family Gyrinidae						
o *Dineutes* sp.	X	X	X			f
*o *Gyrinus* poss. *parcus*	X	X				a
unident. sp.						b
Family Haliplidae						
*ho *Peltodytes festivus*	X	X	X			d
*ho *P.* poss. *muticus*	X	X				a
Suborder Polyphaga						
Family Dryopidae						
ho *Helichus* sp.				X	X	b
Family Elmidae						
o *Heterelmis vulnerata*				X	X	f

TABLE 2.4.   (*Continued*)

Taxon[a]	Substrate[b]					Source[c]
		Lentic		Lotic		
	V	M	S	G	R	
Family Elmidae (*cont.*)						
o *Hexacylloepus ferrugineus*				X	X	f
o *Microcylloepus pusillus*	X	X	X	X	X	f
o *Stenelmis crenata*		X			X	d
*o *Stenelmis* sp.		X			X	a–d
Family Hydrophilidae						
h *Berosus aculeatus*	X	X		X	X	f
h *B. peregrinus*	X	X				c
h *Enochrus* sp.	X	X				f
h *Hydrochus subcupreus*	X	X		X	X	c
o *Paracymus confusus*	X	X	X			f
*ho *Tropisternus lateralis*	X	X	X			a
Family Limnichidae						
o *Lutrochus luteus*				X	X	f
Order Trichoptera						
Family Hydropsychidae						
o *Cheumatopsyche* sp.				X	X	f
*o *Hydropsyche orris*	X		X	X	X	a,f
o *H.* nr. *arinale*	X	X	X	X	X	f
o *H.* nr. *hageni*	X	X	X	X	X	f
*o *Hydropsyche* sp.	X		X	X	X	a
o *Smicridea fasciatella*				X	X	f
Family Hydroptilidae						
h *Oxyethira* sp.	X			X	X	d
Family Leptoceridae						
o *Athripsodes (= Ceraclea?)* sp.	X	X	X	X	X	f
o *Leptocella (= Nectopsyche?)* sp.	X			X	X	b,c,f
h *Oecetis avara*	X		X	X	X	b
h *O. eddlestoni*	X		X	X	X	f
h *O.* nr. *inconspicua*	X		X	X	X	c
h *Oecetis* sp.	X		X	X	X	b,f
unident. sp.						a
Family Polycentropodidae						
Cyrnellus fraternus				X	X	f
C. marginalis		X	X			b
*o *Neureclipsis crepuscularis*			X		X	a,b
o *Neureclipsis* sp.	X				X	f
o unident. sp.						a,d
Order Lepidoptera						
Family Pyralidae						
*h *Nymphula* sp.	X		X			a
Order Diptera						
Suborder Nematocera						
Family Chironomidae						
o *Calopsectra (= Rheotanytarsus?)* nr. *guerla*	X			X	X	d
c *Clinotanypus* sp.	X	X	X			d

(*continued*)

TABLE 2.4.　(*Continued*)

Taxon[a]	V	M (Lentic) S		G (Lotic) R		Source[c]
Family Chironomidae (*cont.*)						
c　*Coelotanypus concinus*	X	X	X			c
o　*Cricotopus bicinctus*	X	X	X	X	X	f
o　*Cricotopus* sp.	X	X	X	X	X	f
o　*Pentaneura* spp.	X	X	X	X	X	c
o　*Polypedilum* prob. *halterale*	X					c
o　*P. illinoense*	X					d,f
Family Chironominae						
Stenochironomus sp.	X					c
Tendipes (= *Chironomus?*) nr. *attenuatus*	X	X	X			d
T. (= *Chironomus?*) *decorus*	X	X	X			c
o　*T.* (= *Dicrotendipes?*) *fumida*	X	X	X			d
o　*T.* (= *Dicrotendipes?*) poss. *modestus*	X	X	X			d
ho　unident. *Chironomini*	X	X	X			c
*ho　unident. spp.						a,b
Family Culicidae						
Anopheles pseudopunctipennis	X	X	X	X	X	b
Family Simuliidae						
o　*Simulium* nr. *vernustum*		X		X	X	f
Suborder Brachycera						
Family Stratiomyidae						
Odontomyia sp.	X	X	X	X		b
Phylum Bryozoa						
Class Ectoprocta						
Order Gymnolaemata						
Family Paludicellidae						
Pottsiella erecta		X		X	X	d
Order Phylactolaemata						
Family Plumatellidae						
Hyalinella punctata				X		e
Plumatella repens	X	X	X	X	X	c,d
Class Endoprocta						
Family Urnatellidae						
*　*Urnatella gracilis*	X					d
Phylum Chordata						
Subphylum Vertebrata						
Class Osteichthyes						
Order Clupeiformes						
Family Clupeidae						
*o　*Dorosoma cepedianum*	X	X		X		a–d,f
o　*D. petenense*				X	X	d,f
Order Lepisosteiformea						
Family Lepisosteidae						
Lapisosteus spp.						
Order Cypriniformes						
Family Catostomidae						
o　*Moxostoma congestum*			X	X	X	b,d
Family Cyprinidae						

TABLE 2.4. (*Continued*)

Taxon[a]	V	Lentic M	Lentic S	Lotic G	Lotic R	Source[c]
o *Cyprinus carpio*		X	X	X		b,d
co *Hybopsis aestivalis marconis*			X	X		b
o *Notemigonus crysoleucas*	X	X				f
Notropis amabilis				X	X	c
N. buchanani	X	X		X		c,d
N. emiliae	X	X				f
*c *N. lutrensis*				X		a–d,f
N. macrostomus				X		b
N. venustus				X		f
*o *Pimephales notatus*	X	X		X		a,b
c *P. vigilax*	X	X		X		c,d,f
Order Siluriformes						
Family Ictaluridae						
o *Ictalurus punctatus*			X	X		f
Pilodictis olivaris						
Order Cyprinodontiformes						
Family Cyprinodontidae						
o *Fundulus notatus*	X	X	X			d,f
Family Poeciliidae						
*o *Gambusia affinis*	X	X				a–d,f
c *Poecilia latipinna*	X	X				b–d,f
Order Atheriniformes						
Family Atherinidae						
c *Menidia beryllina*			X	X		f
Order Perciformes						
Family Centrarchidae						
c *Ambloplites rupestris*	X			X	X	b
Lepomis auritus			X			f
c *L. cyanellus*	X	X				c,d
c *L. gulosus*	X	X	X			f
*o *L. macrochirus*	X	X	X			a–d,f
*c *L. megalotis*			X			a–d,f
c *Micropterus punctulatus*			X	X		d
*c *M. salmoides*	X		X	X		a,b
c *M. treculi*		X	X	X		c
*c *Pomoxis annularis*	X	X				a,d
Family Cichlidae						
ho *Cichlasoma cyanoguttatum*	X	X				c,d,f
Family Percidae						
c *Percina caprodes*			X	X	X	d

*Species found in the habitat described in the section "Structure of Aquatic Ecosystems".

[a]c, Carnivore; co, carnivore-omnivore; dh, detritivore-herbivore; do, detritivore-omnivore (devours organisms and all types of detritus); h, herbivore; hd, herbivore-detritivore; ho, herbivore-omnivore; o, omnivore.

[b]V, vegetation; M, mud; S, sand; G, gravel; R, rock.

[c]a, ANSP (1949); b, ANSP (1951); c, ANSP (1953); d, ANSP (1963); e, ANSP (1967); f, ANSP (1974).

the debris associated with the vegetation were *Amoeba proteus*, *A. radiosa*, and *A. verrucosa*, which are detritivore-omnivores, as are the species of *Bodo*. Other protozoans in and among the debris associated with the vegetation were *Difflugia globosa*, *D. lobostoma*, *Lionotus trichocystis*, and *Oxytricha ludibunda*, which are omnivore-detritivores, as is *Vorticella convallaria*. The latter protozoans are filter feeders.

Several rotifers were in and among the vegetation. Most of the rotifers are omnivores; that is, they feed on small protozoans, bacteria, and algae, particularly diatoms. Those found in and among the vegetation in 1949 were *Euchlanis parva* and *Tripeuchlanis plicata*. Also present were two species belonging to the genus *Lecane*, both omnivores (Table 2.4). *Monostyla cornuta*, *Cephalodella auriculata*, *C. forficula*, and an unidentified species of the genus *Cephalodella* were also present.

On the debris were the gastropods *Amnicola limosa* and *Physa integra*, which is a detritivore-omnivore, as are *Pseudosuccinea columella* and *Helisoma anceps*. The unionid *Amblema plicata* var. *perplicata*, a herbivore-detritivore, was found in and among the plant debris.

Several insects were found in and among the debris. They were the odonates *Hetaerina americana*, *H. titia*, *Hetaerina* sp., *Argia moesta*, *A. translata*, and *Argia* spp., all carnivores. Also present was *Neonura aaroni?*, a carnivore. Other insects present in and among the vegetation were *Gomphus exilis*, *G. externus*, *G.* poss. *militaris*, *G. notatus*, *G. olivaceous*, *Pentagenia vittigera?*, *Tortopus* sp., *Caenis* spp., and *Tricorythodes* spp. Also present was a gerrid, *Ranatra australis*, which is a carnivore. The beetles *Gyrinus* poss. *parcus*, *Peltodytes festivus*, and an unidentified species of the genus *Peltodytes* poss. *muticus* were also present. A beetle belonging to the family Hydrophilidae, *Tropisternus lateralis*, probably a herbivore-omnivore, was present. A few caddisflies were in and among the vegetation: the omnivorous *Hydropsyche orris* and an unidentified species of *Hydropsyche*. One unidentified species belonging to the order Lepidoptera was found. It was an unidentified species of the genus *Nymphula*, a herbivore.

Several fish were found in and among the vegetation: *Dorosoma cepedianum*, an omnivore, and *Pimephales notatus*, an omnivore. Some *Gambusia* were also encountered: *G. affinis*, an omnivore. In the vegetation were the carnivorous *Lepomis macrochirus*, *Micropterus salmoides*, and *Pomoxis annularis*.

Lentic Habitats. The lentic habitats that occurred in pools behind rocks in backwaters and along the edges of the river were a favorite habitat for many species of aquatic life. *Acorus calamus* was present. Several blue-green algae were found in this habitat: an unidentified species of the genus *Cylindrospermum*, *Lyngbya aestuarii*, *Oscillatoria princeps*, *O. splendida*, and *O. tenuis*. Several green algae were found in these lentic habitats: unidentified species of the genus *Oedogonium* and of the genus *Spirogyra*. *Chara gymnopus* was found here. An unidentified species of the yellow-green algae *Vaucheria* was also found in these lentic habitats. Numerous diatoms were found in the pools and along the margins of the river: *Synedra acus*, *Achnanthes affinis*, *A. exigua*, *Cymbella affinis*, *C. cessatii*, *C. hybridiformis*, *C. ventricosa*, *Gomphonema intricatum* var. *vibrio*, *G. parvulum*, *Gyrosigma kuetzingii*, *G. spencerii* var.

nodifera, Navicula cryptocephala var. *pumila, N. gregaria, N. grimmei, N. kotachyi, N. pupula, N. radiosa* var. *tenella, Navicula* spp., *Nitzschia amphibia,* and *N. palea.*

Several protozoans were found in these lentic habitats in 1949. They were the autotrophs *Peranema trichophorum* and *Astasia klebsii.* Several *Euglena* were in these lentic habitats, particularly in the quiet pools: *Euglena pisciformis* and *E. viridis.* Other flagellates found in these habitats were the detritivore-omnivores *Bodo alexeieffii, B. caudatus,* and unidentified species of *Bodo.* Crawling over the debris in these lentic habitats were *Amoeba proteus* and *A. radiosa,* largely detritivores, but they also may eat algae and small protozoans. Attached to the debris was *Actinophrys sol.* Another ciliate in this habitat was an unidentified species belonging to the genus *Prorodon.* In and among the debris in these lentic habitats were specimens of *Paramecium aurelia* and *Oxytricha ludibunda.*

Several rotifers were present in these lentic habitats, that is, in the pools or occasionally in and among the detritus along the margins of the river. They were *Tripeuchlanis plicata, Lecane luna* var. *presumpta, L. stichaea, Monostyla cornuta, M. hamata, M. lunaris,* and *M. pyriformis.* Also present were *Cephalodella auriculata, C. forficula,* and *Cephalodella* sp. *Trichocerca rattus* was also in the pools. Several gastropods were found crawling over the debris in the pools. They were *Physa integra,* which is probably a detritivore-omnivore. Also present in these lentic habitats were several unionid clams. They were *Amblema plicata* var. *perplicata, Lampsilis anodontoides, L. fasciata hydiana, Proptera pupurata, Quadrula aurea, Q. petrina, Q. quadrula* var. *apiculata, Q. verricusa,* and an unidentified species of *Quadrula.* Also found were *Eupera cubensis* and *E. singleyi.* A detritivore-herbivore clam, an unidentified species of the genus *Sphaerium,* was present.

Several insects were found in these lentic habitats: unidentified species of the genus *Hetaerina, Argia moesta, A. translata,* and unidentified species of the genus *Argia,* all carnivores. *Gomphus externus,* a carnivore, was present in the lentic habitats, as were *Gomphus exilis, G.* poss. *militaris, G. notatus,* and *G. olivaceous,* all carnivores. Also present was the carnivorous *Didymops transversa* and an unidentified species of the genus *Macromia.*

A few mayflies were found in these lentic habitats: the omnivorous *Hexagenia bilineata, Pentagenia vittigera?,* and an unidentified species of the genus *Tortopus.* Other mayflies in these lentic habitats were unidentified species of the genus *Caenis,* which are omnivores, and of the genus *Tricorythodes,* also omnivorous species.

A gerrid, *Trepobates inermis,* a carnivore, was in and among the debris, as was *Ranatra australis.* The beetles *Gyrinus* poss. *parcus,* an omnivore, and *Peltodytes* poss. *muticus,* a herbivore-omnivore, were also present. A few elmid beetles were found in these lentic habitats. They were unidentified species of the genus *Stenelmis* and specimens of hydrophilids. Present were the species *Hydropsyche orris* and *Hydropsyche* sp., which are filter feeders and therefore omnivorous, and unidentified species of the same genus. Another trichopteran that was present in these lentic habitats where the current was moderate to very slow was *Neureclipsis crepuscularis.* An unidentified species of the genus *Nymphula,* which is a lepidopteran, was present. It is probably a herbivore.

Several fish were found in the pools: *Dorosoma cepedianum*, an omnivore; *Notropis lutrensis*, a carnivore; *Pimephales notatus*, an omnivore; and *Gambusia affinis*, an omnivore. Also present were *Lepomis macrochirus*, an omnivore, and *L. megalotis*, a carnivore. They were in the pools where the current was moderate. Also present were the carnivorous *Micropterus salmoides* and *Pomoxis annularis*.

Lotic Habitats.   In the lotic habitats were rocks and snags of various types. The current was variable. The substrate was mainly gravel with some sand. Growing attached to the rocks were *Oscillatoria princeps*, *O. splendida*, and *O. tenuis*. A few green algae, *Cladophora glomerata* and *Cladophora* sp., were found on the snags. *Spirogyra* spp. were present where the current was somewhat diminished. An unidentified species of *Vaucheria* was also found in and among the snags.

Several diatoms occurred in this habitat: *Synedra acus*, which was attached to the gravel and snags, *S. ulna*, *S. ulna* var. *danica*, *Achnanthes affinis*, *Cymbella* spp., *Gomphonema lanceolata*, *G. parvulum*, and *Navicula lanceolata*. *Nitzschia frustulum* var. *tenella* was in debris and gravel in between the rocks. In the crevices of the snags was *Nitzschia palea*. The red alga, *Chantransia* sp., was found associated with the rocks and snags.

Among the snags where the current was reduced was the *Amoeba*, *Vahlkampfia limax*, a detritivore. Also present was the ciliate *Nassula ornata*, probably a herbivore-detritivore. A ciliate, *Urocentrum turbo*, was found associated with the snags and gravel. It is probably a detritivore-omnivore, as it ingests its food by cilia. *Vorticella convallaria*, a detritivore-omnivore, was also found in these lotic habitats where the current was reduced, between rocks and in the surface gravel.

Several molluscs were found associated with the rocks. A few unionids were found in and among the gravel. They were *Amblema plicata* var. *perplicata*, a herbivore-detritivore feeding largely on small algae and bacteria. It also probably feeds on some protozoans and would therefore be considered omnivorous. Other bivalves found in this habitat in the gravel were *Lampsilis anodontoides* and *L. fasciata hydiana*. *Proptera purpurata* was also in this habitat and is probably omnivorous. *Quadrula aurea*, which is probably omnivorous or a detritivore-herbivore, was present, as were *Q. petrina*, *Q. quadrula* var. *apiculata*, and *Q. verricusa*. All of these are detritivore-herbivores and eat bacteria, small protozoans, and algae. Other clams were an unidentified species of *Quadrula* and of *Sphaerium*, a detritivore-omnivore.

Several insects were found on the rocks and snags. Associated with the snags was an unidentified species of *Hetaerina*. *Hetaerina* is an odonate and a carnivore. *Argia moesta* is associated with both the rocks and snags and is carnivorous, as are *A. translata* and unidentified species of *Argia*. Several mayflies were found associated with the rocks and snags: *Tortopus* sp. and *Caenis* spp. A few gerrids of the order Hemiptera were also found in this habitat: *Metrobates artus* and *Trepobates inermis*. Another gerrid found in this habitat among the snags and between the rocks was the carnivorous *Rhagovelia choreutes*. The elmid *Stenelmis* sp., an omnivore, was also found in this habitat. A few caddisflies were found in this habitat: the omnivorous *Hydropsyche orris* and an unidentified species of the genus *Hydropsyche*, probably an omnivore. A trichopteran, *Neureclipsis crepuscularis*, was on the snags. It is an omnivorous filter feeder.

In these lotic habitats were the omnivorous *Dorosoma cepedianum*, and in and among the gravel were the carnivorous *Notropis lutrensis* and *Lepomis megalotis*. *Micropterus salmoides*, a carnivore, was also present.

SUMMARY

The Guadalupe River arises in Kerr County, south-central Texas, and flows 712 km in a southeasterly direction, until it empties into the Gulf of Mexico at San Antonio Bay. For the first 260 km the Guadalupe flows through the Edwards Plateau, where the streams forming the headwaters are clear, rocky, and spring-fed. After the Guadalupe leaves the Edwards Plateau it descends the Balcones Escarpment and enters the Texas extension of the Coastal Plain Physiographic Province. Shortly after entering the coastal lowlands, the river contains a number of springs. In this area the Guadalupe flows through rolling-to-flat topography for approximately 170 km, until it is joined by its major tributary, the San Marcos River. This tributary also receives the majority of its flow from groundwater. Following its confluence with the San Marcos, the Guadalupe River is joined by a number of other streams.

The watershed in the Edwards Plateau is dominated by scrub forests of Mexican jupiter, Texas oak, and "live oak." The coastal plain portion of the watershed belongs to the Texas Biotic Province, which is characterized by a moist subhumid climate and deep soils of clay, loam, and sand. Here the climax forest is oak–hickory forest dominated by post oak, blackjack oak, and the hickory oak. Floodplain and riparian assemblages characterized by bald cypress, pecan, elm, and hackberry are common. Whereas land use is limited on the Edwards Plateau, in the lowlands there is considerable agricultural development. In addition, petroleum and natural gas extraction have in the past affected much of the San Marcos subdrainage and the main stem of the Guadalupe. The urban and industrial development is concentrated in the lower Guadalupe River. The biological community described in this chapter is located downstream from the confluence of the Guadalupe and Comal Rivers near Seguin.

In the area of study the shallow-water areas extended out for about 10 m. Most of the channel averaged about 2.4 m in depth in late summer, although many areas were shallower. A moderate-sized, tree-covered island was present in the site of study. The substrate of the river included mud, marl, gravel, and large cobbles and woody debris. In the study area the flow of the river was mainly from the Guadalupe and Comal rivers. A large portion of this flow is from groundwater. Turbidity is often relatively high and the temperature of the Guadalupe River is relatively warm, ranging between 15 and 30°C. However, the surface temperatures near Seguin during the late summer range from 25.6 to 31°C.

The Guadalupe River is a hard alkaline stream dominated by calcium and magnesium hardness and bicarbonate alkalinity. The macronutrient concentrations in late summer near Seguin were relatively high and suggested enriched conditions resulting from agricultural activities. In the upper Guadalupe River above the sewage outfalls near Seguin, the nitrate concentration from the upper Guadalupe was 0.05 mg L^{-1} and that from the Comal River was probably less than 0.02 mg L^{-1}. Downstream from the sewage outfalls nitrate ranged from 0.02 to 0.8 mg L^{-1}. This moderate enrichment resulted in low dissolved oxygen in the summer. However, during most of the year the near-surface dissolved oxygen concentrations were relatively high in late autumn through late winter. It was typically 8 mg L^{-1}, while in the summer the concentrations often range between 4 and 6 mg L^{-1}.

The primary production by benthic algae was relatively high. Algae was relatively abundant, with blue-greens, greens, and diatoms being common. During the summer survey, the algae assemblage was often characterized by an abundance of *Spirogyra* and blue-green algae in 1949, 1950, 1952, and 1962, whereas in 1973 the most abundant algae belonged to the Cladophoraceae and there were few blue-greens or the green alga *Sprirogyra*. This shift may be attributable to the hurricane and associated scouring of muddy habitats. The effects of these hurricane-related floods were also evident in the aquatic macrophytes. The main shift was in the particular species that were present. Another change from pre-1972 and 1973 were the bivalves. The earlier surveys reported a diverse and abundant assemblage of unionids. This assemblage experience a major decline in abundance and a less serious decline in taxonomic richness when the site was surveyed in 1973. This decline was probably the result of intense scouring.

The fish species diversity changed greatly after 1960. The earlier surveys had totals of 8, 13, and 12 species of fish collected at this site near Seguin, whereas in 1962 and 1973 a total of 17 species were collected. The fish fauna was diversified but showed the effects of barriers to upstream movement created by a series of small hydroelectric dams. The fish fauna of the Guadalupe River is greatly influenced by the cooler water of permanent springs. Not only did the springs influence the presence of certain native species but also facilitated the establishment of several introduced species, such as rock bass and the redbreasted sunfish. The moderate-temperature regimes of the large springs have also facilitated the establishment of *Cichlasoma cyanoguttatum* as well as the populations of *Tilapia mossambica* and *T. aurea*. The 1972 hurricanes had profound effects on the benthic fauna and flora. Storm flows in 1972 resulted in considerable deposition of fine sediments These depositional conditions resulted in the accumulation of fine detritus which probably provided a major source of trophic support for benthic macrovertebrates. This is suggested by the abundance of several benthic omnivore-detritivores, including *Hexagenia*, *Caenis*, *Tricorythodes*, chironomids, and oligochaetes. In the Seguin area, the fauna and flora of the Guadalupe River are well developed in sizes of populations as well as in species diversity.

BIBLIOGRAPHY

Academy of Natural Sciences of Philadelphia. 1949. A biological survey of the Guadalupe River at Victoria and San Antonio Bay, Texas. Report for E.I. du Pont de Nemours and Company. ANSP, Philadelphia, Pa. 47 pp.

Academy of Natural Sciences of Philadelphia. 1951. Guadalupe River, Texas. Stream survey report, September 1950. Report for E.I. du Pont de Nemours and Company. ANSP, Philadelphia, Pa. 95 pp.

Academy of Natural Sciences of Philadelphia. 1953. Guadalupe River, Texas. Stream survey report for E.I. du Pont de Nemours and Company, September 1952. ANSP, Philadelphia, Pa. 83 pp.

Academy of Natural Sciences of Philadelphia. 1963. Guadalupe River, Texas. Stream survey report for E.I. du Pont de Nemours and Company. ANSP, Philadelphia, Pa. 67 pp.

Academy of Natural Sciences of Philadelphia. 1967. Cursory stream survey report of the Guadalupe River for the E.I. du Pont de Nemours and Company, 1966. ANSP, Philadelphia, Pa. 21 pp.

Academy of Natural Sciences of Philadelphia. 1974. Guadalupe River survey 1973 for the E.I. du Pont de Nemours and Company. ANSP, Philadelphia, Pa. 100 pp.

Blair, W. F. 1950. The biotic provinces of Texas. Tex. J. Sci. 2: 93–117.

Brown, W. H. 1953. Introduced fish species of the Guadalupe River basin. Tex. J. Sci. 5(2): 245–251.

Epperson, C. R., and R. A. Short. 1987. Annual production of *Corydalus cornutus* (Megaloptera) in the Guadalupe River, Texas. Am. Midl. Nat. 118(2): 433–438.

Fenneman, N. M. 1931. Physiography of western United States. McGraw-Hill, New York. 534 pp.

Fenneman, N. M. 1938. Physiography of eastern United States. McGraw-Hill, New York. 714 pp.

Hannan, H. H., and L. Broz. 1976. The influence of a deep-storage and an underground reservoir on the physicochemical limnology of a permanent central Texas river. Hydrobiologia 51(1): 43–63.

Hannan, H. H., W. C. Young, and J. J. Mahew. 1972. Nitrogen and phosphorus dynamics in three central Texas impoundments. Hydrobiologia 40(1): 121–129.

Hannan, H. H., W. C. Young, and J. J. Mahew. 1973. Nitrogen and phosphorus in a stretch of the Guadalupe River, Texas, with five main-stream impoundments. Hydrobiologia 41(3): 419–441.

Hubbs, C. 1957. Distribution patterns of Texas fresh-water fishes. Southwest. Nat. 2(2–3): 89–104.

Hubbs, C., R. A. Kuhne, and J. C. Ball. 1953. The fishes of the upper Guadalupe River, Texas. Tex. J. Sci. 10(4): 452–483.

Lee, D. C., C. R. Gilbert, C. H. Hocutt, R. E. Jenkins, D. E. McAllister, and J. R. Stauffer, Jr. 1980. Atlas of North American freshwater fishes. North Carolina State Museum of Natural History, Raleigh, N.C. 854 pp.

Pflieger, W. L. 1975. The fishes of Missouri. Missouri Department of Conservation, Jefferson City, Mo. 343 pp.

Stanley, E. H., R. A. Short, J. W. Harrison, R. Hall, and R. C. Wiedenfeld. 1990. Variation in nutrient limitation of lotic and lentic algal communities in a Texas (USA) river. Hydrobiologia 206(1): 61–71.

U.S. Geological Survey. 1961. Surface water supply of the United States, 1960. Part 8. Western Gulf of Mexico basins. Water-Supply Pap. 1712. USGS, Washington, D.C.

U.S. Geological Survey. 1970. Surface water supply of the United States, 1961–1965. Part 8. Western Gulf of Mexico basins. Vol. 2. Basins from Lavaca River to Rio Grande. Water-Supply Pap. 1923. USGS, Washington, D.C.

Visher, S. S. 1954. Climatic atlas of the United States. Harvard Univ. Press, Cambridge, Mass. 403 pp.

Young, W. C., and C. W. Bayer. 1979. The dragonfly nymphs (Odonata: Anisoptera) of the Guadalupe River basin, Texas. Tex. J. Sci 31(1): 85–97.

Young, W. C., H. H. Hannan, and J. W. Tatum. 1972. The physico-chemical limnology of a stretch of the Guadalupe River, Texas, with five mainstream impoundments. Hydrobiologia 40: 297–319.

CHAPTER *3*

Rio Grande

INTRODUCTION—DESCRIPTION OF THE WATERSHED

The Rio Grande drainage basin forms one of the most important international waterways in the United States. It is rich in culture and history as well as in controversy over use and management of the water of the Rio Grande. The Rio Grande basin above El Paso, Texas is one of the oldest regions of agriculture in the United States. Agricultural activities extend back centuries to prehistoric inhabitants of the Rio Grande valley, and include the seventeenth and eighteenth century Pueblo Indians and Spanish colonists and, in the latter part of the nineteenth century, European-Americans (Wozniak, 1987). More recent history involves disputes and concerns over flood management, irrigation, and distribution and delivery of upstream waters to downstream users in an attempt at fair sharing among concerned parties. Because of the long history of agricultural activity, Rio Grande water is tied to public laws governing its conveyance, storage, and use. The close connection between legislation and flow of water through the Rio Grande is largely responsible for the present physical state of the river, floodplain, and associated riparian communities. Changes in the floodplain ecology probably began shortly after human settlement in the region, and change continued relatively unabated with increasing population (Bullard and Wells, 1992). The Rio Grande watershed is the sixth largest watershed in North America (Hansmann and Scott, 1977).

GEOLOGY

The geological history of the Rio Grande watershed dates from the Late Cretaceous Period. It involves several geological events, including warping and uplifting of Earth's crust, volcanic eruption, erosion, and sedimentation. Seas covered this region of New Mexico during the Cretaceous Period and receded to form a large floodplain. Through compression of Earth's crust, a linear upward development along the entire New Mexican Rocky Mountain axis took place, which eventually gave rise to the mountain range bordering the Rio Grande watershed. These uplifts were paralleled by downwarping, forming a series of parallel basins which, over thousands of years, have accumulated tremendous quantities of material carried downstream by the Rio Grande. The river also contains erosion material from the surrounding mountains (Hansmann and Scott, 1977). Records for the last 100 years show that the Socorro-Albuquerque sections have experienced considerable, but rather localized, earthquake activity (Hansmann and Scott, 1977).

CLIMATE

The Rio Grande drainage basin is located in a transitional climatic zone between the Gulf of Mexico and the Pacific rainfall provinces. As in other parts of the southwestern part of the United States, the Rio Grande basin has an arid to semiarid climate, characterized by abundant sunshine. It has relatively low humidity. There are wide fluctuations in the diurnal temperature. The drainage basin upstream from Bernalillo has a semiarid climate, whereas downstream from Belen the climate is arid. A temperate climate is present in the high mountain ranges (Bullard and Wells, 1992).

Precipitation
The annual precipitation averages from 178 to 380 mm over two-thirds of the basin. Only in the highest mountain areas does it exceed 635 mm. Occasionally, snow is present on the ground at low elevations for more than 24 hours, but the winters are generally dry. In the higher mountains, snowfall comprises 30 to 75% of the annual precipitation. In the rest of the basin, snowfall comprises less than 25% of the annual precipitation. Over half of the annual moisture is supplied by summer precipitation, and most of the rainfall is brief. However, some thunderstorms are intense and convective. Summer thunderstorms have a moderating effect on the temperature during the day. During the spring months, winds are continuous and the prevailing winds are from the southwest. Throughout the lower elevations of the basin, evaporation is high, and in the southern part of the basin where arid conditions exist, the evaporation rate is highest (Bullard and Wells, 1992).

The upper Rio Grande in Colorado and New Mexico and the Rio Conchos in Chihuahua receive flow from snowmelt (Tamayo and West, 1964); springs, seasonal rains, and occasional tropical storms provide most of the flow in the lower part of the basin. In Texas, the upper part of the basin has less precipitation than the lower part. Rainfall averages 7.82 in. at El Paso, 12.21 in. at Fort Stockton, 20.14 in. at Laredo, and 25.44 in. at Brownsville (Bomar, 1983; [Texas Natural Resources Conservation Commission (TNRCC), 1994].

Storms

The source of the major part of the precipitation in the basin is the tropical region of the Gulf of Mexico (Jetton and Kirby, 1970; Tuan et al., 1973). Significant amounts of moist unstable air imported by hurricane activity on the west coast of Mexico may result in major storms. The polar Pacific and polar continental air masses influence the position of the jet stream. This is important, as they affect the magnitude and location of precipitation in the drainage basin (Bullard and Wells, 1992).

There are two types of storms in the region. Local thunderstorms may be the result of orographic or convective lifting of water, whereas frontal storms are caused by the interaction of two or more air masses. Typically, a storm period lasts less than 24 hours, although the intensity of the precipitation may be extremely high at some locations within the general area of the storm. Tropical disturbances, which are related to hurricane activity in the Gulf of Mexico or in the Pacific Ocean off the west coast of Mexico, may cause the precipitation periods to last more than 24 hours (Bullard and Wells, 1992).

The type and magnitude of the storms are typically seasonal. For example, the summer months (June through August) characteristically have intermittent importations of tropical marine air masses that cause the storms. Isolated shower activity may result from orographic lifting or convective action. These showers are often intense but usually localized. This type of activity may occur at many points simultaneously and may infrequently cause general storms (Bullard and Wells, 1992).

During the winter months (November through February) vigorous air-mass activity may occur. These vigorous air masses are characterized by cold fronts of polar Pacific air that move eastward and northeastward (Maker et al., 1972). In the mountain ranges, precipitation occurs as snow, whereas in the lower elevations of the central basin, the precipitation may be moderate rains or snow. Precipitation in the southern part of the basin is generally low because of the northerly path of storms (Bullard and Wells, 1992).

The periods transitional to summer and winter (March through May and September through October) are associated with the largest storms that generate floods. Greater temperature differences between air masses cause an increase in air-mass instability. Various types of air masses move simultaneously into the region. These are more pronounced during transition from summer to winter than during other periods. In the mountains, in conjunction with general rains elsewhere in the basin, heavy snowfall may occur. Storm activity during later spring months and the fall often is characterized by high-intensity rains over large portions of the area (Bullard and Wells, 1992). High-intensity rainfall events during July 1988 caused considerable flood damage on the piedmont slope in and around Albuquerque (Bullard and Wells, 1992).

Temperature

In high altitudes where dry continental climates prevail, the average temperature ranges from a minimum of 2.9 to 8.3°C to a maximum of about 23 to 24°C along the river valley. Mean annual temperature ranges from about 12.2 to 16.1°C. In the high mountain elevations, freezing temperatures sometimes occur. In these upper basin regions and along the drainage divides, the average summer temperatures may be 5°C lower than in the valley regions. The frost-free period is from May to October in the valley regions. July is the hottest month in the valley; the lowest temperatures occur in January (Bullard and Wells, 1992).

Runoff

Spring snowmelt and spring and summer convective thunderstorms are largely the cause of runoff in the basin. In the mountains precipitation ranges from less than 25 mm to greater than 255 mm. About 70% of the runoff occurs from May to August, during snowmelt and thunderstorm activity (Bullard and Wells, 1992).

VEGETATION OF THE WATERSHED

Woodland Biome

The woodland biome is found in elevations from about 4500 ft to 7500 ft above sea level. It is present in the foothills and lower parts of the mountains, including the Sandia, Manzano, San Mateo, Jemez, and other mountains of the Rio Grande watershed. The misuse of grasslands has led to the extension of this biome into the lower elevations. The woodland usually consists of dominant species of one-seed juniper, piñon pine, and several species of oak. Several species of shrubs and herbaceous plants are also associated with this biome. Grasses are scattered and contain species of the mixed grassland association in and among the trees and shrubs. The forest consists of some spruce, Douglas fir, juniper, and piñon pines (Hansmann and Scott, 1977). Shrubs are sagebrush, creosote bush, saltbush (commonly called black greasewood), and juniper. The immediate vegetation surrounding the Rio Grande river is that of the grassland associations.

Mixed Grassland Association

The mixed grassland association consists of midheight grasses and short grasses in equal amounts. It is found on the plains of the lower elevations and occupies a large part of the rolling plains that border the lower desert grasslands through which the Rio Grande flows. Some variation exists in this association. Blue grama, hairy grama, and galleta grass are the most common grasses. Several dominant herbaceous plants and shrubs are present, such as locoweed, milkweed, thistle, saltbush, snakeweed, and greasewood. Because of overgrazing and drought, there has been deterioration of this association, and shrubs such as snakeweed, rabbitbrush, shadscale, and saltbush are rapidly invading the area (Hansmann and Scott, 1977). At higher elevations, trees and other shrubs are more dominant.

Desert Grassland Association
The desert grassland association extends across the southern part of New Mexico, with northward extensions into the major river valleys. In the Rio Grande watershed of New Mexico, it covers the southern portion of the watershed and extends up the middle of the Rio Grande Valley as far north as approximately the town of Bernalillo. It ranges in altitude from 2850 to 5500 ft above sea level. A striking feature of this association is the dominance of shrubs, such as creosote bush, tarbush, mesquite, and others. This dominance of shrubs has been interpreted as an area known as the desert shrub association. However, historically, this area was originally grass covered, and the shrub development was caused by deterioration due to overgrazing and periodic drought (Hansmann and Scott, 1977).

HISTORICAL DESCRIPTION

Records kept by early explorers and surveyors of the Rio Grande valley describe an environmental setting and a river very different from that today. The width of the channel varied from 300 to 600 ft (90 to 180 m) except where it was constricted in narrow hard-wall canyons. For most of the year the invariably turbid waters occupied only a portion of the riverbed. There were extreme seasonal variations in flow: low in the winter months, sometimes nonexistent in the summer, and often above flood stages during the late spring snowmelts in the headwaters. A torrential summer rainstorm could transform the river abruptly from a trickle into a torrent, transporting a large quantity of sediments washed from the highly erodible valley soils carried through the usually dry arroyos.

During floods the river often cut new channels, abandoning entire sections. Oxbow lakes and marshy areas were not uncommon. The watershed described by Bartlett was generally sparsely vegetated except for fine stands of cottonwoods and mesquite whenever the road approached the river. Patches of grass and shrubs were occasionally encountered along the route. The sparseness of the vegetation, however, was due only to the shortage of water, for the soil of the bottomlands was exceedingly productive when water was provided.

Agriculture has existed on the watershed almost continuously, longer than at any other place in the United States, beginning with Pueblo Indians, who irrigated and farmed the bottomlands for almost 500 years before the arrival of Coronado in 1540. The Spanish occupation of New Mexico began in 1598, but there were no major population increases until the influx of Anglo-American settlers in the latter part of the nineteenth century. Agriculture was the main means of livelihood, and populations centered around irrigated lands along the river. In recent years there has been a diversion from agriculture to urban land uses because of heightened population.

GENERAL DESCRIPTION OF THE RIO GRANDE RIVER

Colorado
The Rio Grande headwaters lie along the Continental Divide at elevations ranging from 2440 to 3660 m in the San Juan Mountains of southern Colorado. The Rio

Grande drains approximately 3370 km^2 of these high mountain peaks in the north-ernmost part of the basin (Bullard and Wells, 1992). In this area the snowpack may accumulate to a depth of 50 ft (Hansmann and Scott, 1977).

The San Juan Mountains to Elephant Butte Dam

The drainage basin area above Elephant Butte is about 76,275 km^2, including 7,615 km^2 in the Closed Basin of the San Luis valley in Colorado. Above Velarde the drainage basin area is about 27,325 km^2. The Rio Chama, one of the most important tributaries in the study area, has its headwaters in the Jemez, Conejos, and San Juan Mountains of New Mexico and Colorado. The Rio Chama drainage basin is about 8160 km^2. Other important tributaries are the Rio Puerco and the Rio San Jose (Bullard and Wells, 1992).

The valley floor and plains of the Rio Grande in New Mexico have been farmed for hundreds of years, with the Rio Grande supplying the needed water. The watershed drains approximately $\frac{1}{4}$ million square miles of land. A large part of the resulting water is derived from mountainous areas of southern Colorado and northern New Mexico (Hansmann and Scott, 1977).

Elephant Butte to El Paso

In the segment of the Rio Grande between Elephant Butte and El Paso is the Caballo Dam. Flow has been modified considerably below the Elephant Butte Dam and the Caballo Dam. Releases from the dams have been according to needs. In the winter no water is released from Caballo Dam, and the flow below the dam depends largely on base flow, occasional rainfall, and municipal wastewater discharges.

El Paso to Fort Quitman

The segment of the Rio Grande between El Paso and Fort Quitman consists largely of tailwater supplied from the Elephant Butte Dam. In this segment the annual rain-fall averages 200 mm (7.8 in.), the lowest in Texas.

Fort Quitman to Amistad Reservoir

Below Fort Quitman, the Rio Conchos from Mexico is the largest entry into the Rio Grande. It is located 454 km (2.84 mi) south of El Paso and enters the Rio Grande just below Presidio. This flow continues to Amistad Dam, located 500 km (312 mi) below Presidio. No major tributary flows into the Rio Grande from the U.S. side until the in-flow of the Pecos River at Langtry and the Devil's River at Amistad Reservoir. The annual rainfall in this section of the Rio Grande averages 250 to 300 mm (10 to 12 in.).

The Fort Quitman to Amistad Dam segment has two inflows from the U.S. side and the Rio Conchos from the Mexican side. The Rio Conchos accounts for 46% of the recorded inflow, the Devil's River 22%, and the Pecos River 17% in this segment of the Rio Grande. There is a net increase in inflow in the Rio Grande between Pre-sidio and Amistad Dam of 284 million cubic meters which is not accounted for by these record inflows. The unaccounted-for flow was divided in proportion to the drainage areas for the Texas side, 20,000 km², and the Mexican side, 15,600 km². Between Fort Quitman, Colonia Luis Leon, and Amistad, the total annual inflow from the U.S. side was estimated to be 861 million cubic meters and that from the Mexican side, 1033 million cubic meters in this section of the Rio Grande.

Amistad Dam to Falcon Reservoir
The Rio Grande between Amistad Dam and Falcon Reservoir is a long stretch extending 481 km (299 mi). There is no major tributary in this reach, but there are numerous creeks and draws that flow into the Rio Grande after storms. The annual rainfall of this section increases to 500 mm (20 in.) per year.

The Amistad–Falcon segment starts with the inflow of Arroyo de Los Jaboncillos (four springs and three creeks near Ciudad Acuña) from the Mexican side followed by the inflow of four Mexican rivers, including the Rio Salado. The record total surface inflow from the Mexican side amounts to 1.01 billion cubic meters annually in this segment of the Rio Grande, and the Rio Salado accounts for 47% of the inflow. The recorded inflow from the Texas side, which includes irrigation return flow from the Maverick Irrigation District, amounts to 374 million cubic meters. Municipal sewage from Eagle Pass and Laredo provides an additional inflow of 12 million cubic meters per year. Sewage water is also discharged from the Mexican side into the Rio Grande: that is, from Nuevo Laredo. The exact quantities are unknown but are probably comparably small. The Rio Grande gains flow from the Amistad and Falcon dams of 983 million cubic meters. The net diversion at the Maverick power plant is 288 million cubic meters, which is then channeled into the Maverick irrigation ditches. Additional diversion to Eagle Pass and Laredo is estimated at 12 million cubic meters. The diversion in Mexico is not recorded but is an estimated 26 million cubic meters based on irrigated acreage. The gain in flow plus the quantities diverted is estimated at 1.31 billion cubic meters, which approximately equals the estimated total inflow of 1.4 trillion cubic meters per year. Seventy-two percent of the inflow in this region of the Rio Grande originates from the Mexican side.

Falcon Reservoir to Mouth
The Rio Grande below Falcon Reservoir to the Gulf of Mexico extends 442 km (275 mi). The Rio Salado from Mexico is the major tributary that flows directly into the Falcon Reservoir, and the Rio San Juan flows into the Rio Grande below Falcon. There are two major drainways on the U.S. side, the main floodway and the Arroyo Colorado. The latter is of special importance because it flows directly into the Laguna Atascosa National Wildlife Refuge.

The Falcon to the Gulf coast segment has a topographical slope where a large portion of the Rio Grande riverbed is higher than the elevation of the drainage basin on the Texas side. The general direction of surface flow is toward the Laguna Atascosa and the Laguna Madre, away from the Rio Grande. The inflow into the Rio Grande is thus from the Mexican side (chiefly from the Rio San Juan and the San Juan drainage) and is recorded to be 628 million cubic meters annually. The reduction in flow of the Rio Grande between Falcon Dam and Brownsville averages 1865 billion cubic meters, while the record plus some estimated diversion anamolies is 2477 billion cubic meters annually. The recorded diversion exceeds the total true inflow (637 million cubic meters) by 1.84 billion cubic meters, which coincides with the measured reduction in flow. Overall, the recorded surface inflow on the Texas side amounts to 1.835 billion cubic meters, and that from the Mexican side, 2.67 billion cubic meters annually, which is roughly a 1:1.5 ratio in favor of the Mexican side. This ratio, however, excludes surface inflow into the Rio Grande, water use, and irrigation use.

The natural drainage flow is away from the Rio Grande eastward toward the Laguna. This area is outside the Rio Grande basin and is part of the Nueces River coastal basin.

RIVER COURSE

The Rio Grande originates in the San Juan Mountains of southwestern Colorado and flows south through New Mexico and along the U.S.-Mexican border (Figure 3.1) for some 3036 km before it reaches the Gulf of Mexico (Sublette et al., 1990).

CHARACTERISTICS OF THE RIO GRANDE AND ITS WATERSHED

Colorado

In the headwaters of the Rio Grande snowpack may accumulate to a depth of 50 ft. Runoff in this area continues throughout the spring and the summer. Heavy snowmelt is usually confined to a period of about 40 days. However, in this period of time, after an average year of snowfall, an estimated 500 million tons of water is passed into the main stream. Runoff may exceed 1 billion tons of water after an exceptionally wet winter (Reynolds, 1975). This runoff is partially controlled by a dam north of the San Luis valley of southern Colorado. The lake resulting from this dam has a capacity of 50,000 acre-ft. Because of downstream agricultural demands, the water level varies during the summer months. After leaving the dam, in the next 50 mi the Rio Grande drops about 3000 ft in elevation. The river consists of steep rapids and deep pools. It is an excellent habitat for rainbow trout (Hansmann and Scott, 1977).

As the river enters the San Luis valley of the Colorado Plateau (known as the San Luis basin), the velocity of the water slows down. This occurs about 8 mi below Creede, Colorado. This valley is the first area where agriculture takes place. Much of this water, enriched with salts and silt, returned to the main channel (Hansmann and Scott, 1977).

New Mexico

The river enters New Mexico west of the town of Costilla in Taos County (Figure 3.2). The elevation is about 7400 ft above sea level. This watershed is part of the Taos Plateau and is arid. As the river runs through this plateau region, the water velocity increases and the river forms a deep wedge through the volcanic basalt and ash. The resulting river channel forms a deep irregular canyon with boiling rapids of water. It

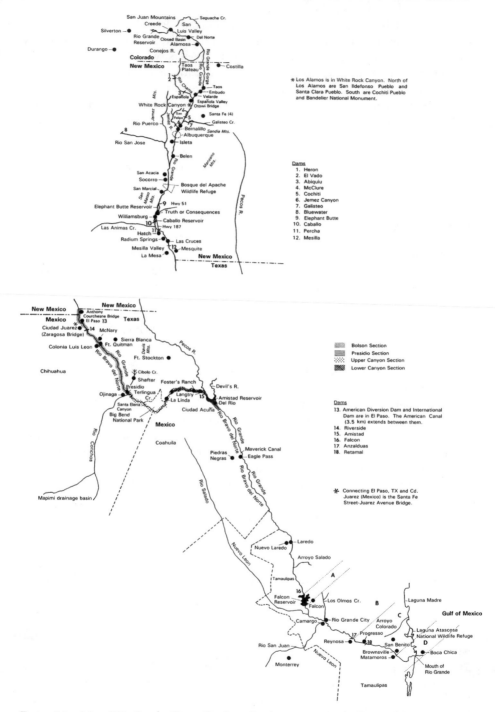

FIGURE 3.1.   Map of Rio Grande. (Created by Susan Durdu; drawn by Su-Ing Yong.) "Rio Bravo del Norte" is another name for the Rio Grande.

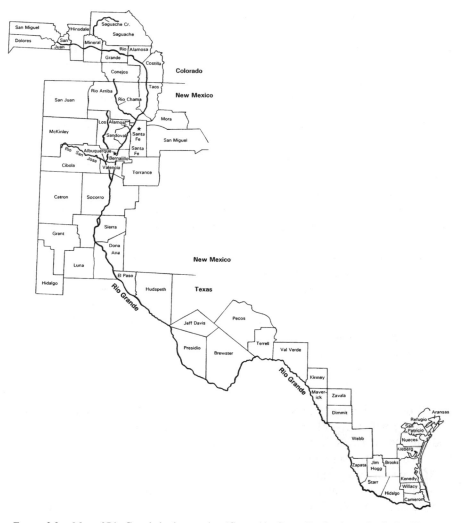

FIGURE 3.2.   Map of Rio Grande basin counties. (Created by Susan Durdu; drawn by Su-Ing Yong.)

is approximately 305 m deep with a floodplain typically no wider than 15 m. It is referred to as the "Rio Grande Gorge." The river leaves the canyon at Embudo, New
Mexico. The water velocity slows down south of this area as the Rio Grande flows
into the Española valley (known as the Española basin). In this valley the water is
funneled through the irrigation cycle of diversion and return channels. After it
leaves the valley, the Rio Grande increases velocity for the next 25 mi, as it passes
through the gorge known as White Rock Canyon. This area marks the southern limits of the relatively fast moving waters in the Rio Grande in New Mexico (Hansmann
and Scott, 1977).

The river continues south through a broad valley that extends to Otowi Bridge (just south of Santa Clara Pueblo), then becomes abruptly constricted as it enters another canyon that extends downstream to Cochiti Pueblo. The remainder of the river's course in New Mexico is through a broad valley of low relief. Stream gradients of 4.75 m km^{-1} are common upstream of Cochiti Pueblo, with stream gradients of 0.95 m km^{-1} prevailing between Cochiti Pueblo and El Paso, Texas (Sublette et al., 1990).

The Rio Grande upstream of Cochiti Pueblo is perennial with numerous tributaries. Downstream, much of the river may be dewatered for extensive periods. Prior to settlement of the Rio Grande valley by Anglo-Europeans, evidence indicates that the river was perennial throughout its course in New Mexico, with even the most arid portions of the floodplain characterized by numerous marshes, oxbow pools, and a fringe forest of cottonwoods, willows, and shrubby phreatophytes (Metcalf, 1967).

In this reach it is impounded in the Cochiti Reservoir behind a massive earth-filled dam which was developed to control the sediment load. The gradient is relatively small in this valley. It is approximately 5 ft/mi or less. The river meanders through the heavily populated areas of Bernalillo, Albuquerque (Albuquerque basin), Belen, and Socorro. The stream channel has been manipulated in this area and thus forms oxbow lakes and marshlands. Many of these marshlands are federal or state wildlife refuges. In many areas irrigation diversion has depleted the surface waters of the main channel (Hansmann and Scott, 1977). After leaving Elephant Butte and Caballo reservoirs, the river is funneled into a network of channels that irrigate the Mesilla valley. This area is one of the most productive expanses of farmland in New Mexico.

Texas
The Rio Grande enters Texas near El Paso. At this point the elevation is 3700 ft. From this point the river forms the International Boundary between the United States and Mexico. In Texas the water is impounded in several locations, and large quantities are diverted for agricultural uses (Hansmann and Scott, 1977).

After traveling approximately 1900 mi, with a drop in elevation of approximately 2 mi from its headwaters, the inorganic and organic silt-laden waters of the Rio Grande enter the Gulf of Mexico at Brownsville, Texas (Hansmann and Scott, 1977).

FLOW OF THE RIVER

Surface Flow
The headwaters are some 12,000 ft above sea level, and in the upper part the flow is rapid as it descends in mountainous canyons to San Luis valley in Colorado. From there the flow of the river has a moderate speed through broad valleys as it enters a long canyon with a maximum depth of about 400 ft which extends to about 4 miles above the boundary between Colorado and New Mexico. On its course in New Mexico it is hemmed in by canyon walls that may be 1000 ft high, or it may flow through narrow mountain valleys. It passes through a series of picturesque canyons in the Big Bend, some of them 1750 ft in depth in the Big Bend and becomes a silt-laden stream

with shifting channel as it flows through Texas. This is the area between the San Luis valley and the mouth of the Conchos (Finch et al., 1998).

The flow of the Conchos is constant and, in the Big Bend, the volume of the Rio Grande is enhanced by springs that break out in the bed. The flow of the Rio Grande is 10 times higher in some years than in others, and great floods have been known to occur in the lower course. On its course through the Coastal Plains, the channel of the Rio Grande is greatly obstructed by sandbars, and thus the river is of little importance to navigation (Finch et al., 1998).

Groundwater Flow
The groundwater flow of the Rio Grande is best understood in the middle section of the Rio Grande, that is, within the middle Rio Grande valley. Here the principal aquifer system is composed of Tertiary sediments of the Santa Fe Group and Quaternary deposits. These sediments are the main rift-filling sediments, consisting dominantly of clays, silts, gravel, and sand (Bullard and Wells, 1992).

Recharge of the principal aquifer is mainly from two sources. They are the margins of the basin and the Rio Grande valley. Small aquifers are formed by Paleozoic and Mesozoic sedimentary rock units. These are exposed along the margins of the Rio Grande Rift. The flow in these rocks is toward the Rio Grande. The small aquifer system is recharged through infiltration in the bed material of arroyos that intersect with the bedrock units of the aquifer system (Bullard and Wells, 1992).

The recharge of groundwater is mainly from the Rio Grande proper. Groundwater flow in the valley is controlled by the river, conveyance channels, acequias, ditches, laterals, drains, and by groundwater from adjacent areas. The floodplain may contribute to shallow alluvial aquifer recharge. Shallow-water aquifers are also recharged from irrigation infiltration. Also contributing are drains, which were designed and constructed so that they discharge excess waters into the aquifers and thus prevent waterlogged soils. This promotes flushing of the soils of salts that tend to be concentrated by evaporation. These drains also intercept regional groundwater inflow (Bullard and Wells, 1992).

These drains are the main factor in controlling the level of groundwater in the Rio Grande valley. Therefore, groundwater withdrawals by irrigation or domestic use are quickly replaced by return flow of irrigation water. These return-flow irrigation waters include infiltration of surface waters from acequias, ditches, laterals, the river, the riverside drain, and by conveyance channels or drains (Bullard and Wells, 1992).

Peter (1987) made a short study of the groundwater flows and shallow aquifer properties. This study indicated that groundwater levels usually fluctuate on a seasonal basis, thus being lower in the summer months (when irrigation and groundwater withdrawals are greater) and higher during winter months (Bullard and Wells, 1992).

The quality of the shallow groundwater is greatly affected by the irrigation of the floodplain. It is evident from studies that groundwater has been contaminated by nitrates and organic compounds such as gasoline and diesel fuels. The infiltration of nitrates is to be expected because of agricultural practices. Studies indicate that the excessively high chloride concentration in the Rio Grande is influenced by the principal aquifer system, which brings deep, chloride-enriched waters into contact with the near-surface aquifer system. It is probable that the groundwater in other sections of the Rio Grande has a profound effect on the surface waters (Bullard and Wells, 1992).

Contributions of Tributaries to Flow
Contribution of Pecos and Devil's Rivers. The Pecos River and Devil's River contribute 274 to 353 million cubic meters annually to the flow of the Rio Grande, respectively. All of these flows are stored at Amistad International Reservoir. The discharge from Amistad Dam has averaged 2.06 billion cubic meters annually since its construction in 1968. About half of this release is taken into the Maverick Canal located 28 km south of Del Rio for hydraulic power generation and irrigation. The return flow goes back into the Rio Grande, and the remainder is used for irrigation throughout the Maverick Extension Canal. The combination of the base flow, return flow, and inflow from the creeks brings the flow of the Rio Grande back to over 2 billion cubic meters annually at Eagle Pass. The diversion below Eagle Pass but above Laredo is minimal, and the Rio Grande gains flow and reaches 2.8 billion cubic meters annually at Laredo.

Contribution to Flow below Laredo. Below Laredo there are several rivers and streams that flow into the Rio Grande. The Rio Salado from Mexico is one of the largest rivers and has contributed to the flow of the Rio Grande at an annual rate of 472 million cubic meters. The combined flow reaches 3.0 billion cubic meters annually at Falcon International Reservoir. Below Falcon, the Rio San Juan (434 million cubic meters per year) flows into the Rio Grande from the Mexican side at Camargo. The Rio Grande water is diverted between Rio Grande City and Anzalduas Dam at a rate of 292 million cubic meters per year for irrigation. The major diversion in Mexico is the Reynosa.

The U.S. side of the diversions are the Anzalduas Dam, town of Progresso, and town of San Benito, at a combined diversion flow of 919 million cubic meters per year. When the Rio Grande reaches Brownsville the flow decreases to 1.18 billion cubic meters per year, which includes erratic floodwater after a storm. There is no recorded inflow from the Mexican side in this segment of the Rio Grande.

Factors Affecting Flow
The pattern of flow in the Rio Grande is highly variable depending on the structure of the watershed and the bed material over which it flows. In some reaches the flow is fast and uniform, and in other reaches, braided channels exist. Sediment load may vary greatly depending on the region through which the river flows and the structure of the channel. The channel pattern may change seasonally depending on the amount of flow and the sediment load (Bullard and Wells, 1992).

Typical characteristics of flow of the Rio Grande and its tributaries vary according to the position in the drainage basin. This is a reflection of the geology and topography of the physiographic regions traversed by the river. Throughout the length of the river, the gradient, channel pattern and width, discharge, and sediment loads are variable (Bullard and Wells, 1992).

River Drainage Area. The entire area of the Rio Grande drainage basin is about 470,000 km^2, of which 230,000 km^2, are in the United States and the remainder in Mexico. The river flows south from Colorado through the length of New Mexico and then forms the international boundary between Texas and Mexico along its 3220-km route to the Gulf of Mexico (Bullard and Wells, 1992).

Gradient. The gradient of the Rio Grande is high in the headwaters as it flows through the steep canyons on its way to the San Luis valley. Locally, the gradient may be tens of meters per kilometer. Through the San Luis valley, the river has a gradient of about 0.56 m km^{-1}, whereas through the Rio Grande Gorge, the gradient ranges from 2.25 to greater than 28.4 m km^{-1}. From Velarde to Cochiti Reservoir the gradient is 1.9 m km^{-1}. Below the Cochiti Dam to Elephant Butte the gradient is much less, being about 0.76 m km^{-1}. Below the Elephant Butte Dam, the gradient is relatively low (Bullard and Wells, 1992).

Characteristics of the Rio Grande Channel. The Rio Grande channel varies dramatically according to the geographic location within the river basin. Characteristics of the channel such as width and sinuosity are greatly influenced by the position within the drainage basin and the proximity of tributaries that discharge large volumes of sediments into the main stem. The width of the Rio Grande valley varies from less than 200 m in the Rio Grande Gorge to 1.5 to 10 km from Velarde to Elephant Butte with the exception of White Rock Canyon and the San Marcial Constriction. The floodplain is very variable in the Rio Grande, depending on its location (Bullard and Wells, 1992).

Sediment Load of the Rio Grande and Tributaries

The suspended sediment loads in the Rio Grande and its tributaries are variable. These are to a certain extent regulated by flood and sediment control structures, particularly in regions above Albuquerque. An increase in the sediment load can have a dramatic effect on the behavior of the river and its geomorphology both upstream and downstream from where this occurs.

At Otowi Bridge, in 1987 the maximum daily suspended sediment load was 23,000 tons [U.S. Geological Survey (USGS), 1987], whereas in 1961 the maximum daily suspended sediment load was 332,000 tons. Downstream at San Acacia (225 km from Otowi Bridge), the maximum suspended sediment load recorded daily was 1,597,000 tons in 1961. This had been reduced greatly by 1987, when the maximum and minimum daily suspended loads were 180,000 and 5.4 tons, respectively (Bullard and Wells, 1992).

Influence of main-stem impoundments and tributary contributions on sediment load is indicated by data for the Rio Grande just below Cochiti Dam. The maximum daily recorded suspended sediment load was 4580 tons and the minimum was 0.02 ton at Cochiti Dam. In contrast, by 1987, the maximum and minimum suspended sediment loads were 442 and 15 tons, respectively. In the 40-km reach between Otowi Bridge and Cochiti Dam, there is daily deposition of thousands of tons of sediment which indicates that the upper basin sediments are being effectively trapped by the Cochiti Reservoir. The same situation exists for sediment transported by tributaries of the Rio Chama and trapped in reservoirs on the Rio Chama (Bullard and Wells, 1992).

The major tributaries of the Rio Grande are the Rio Chama, Jemez River, Rio Puerco, Rio Conchos, Pecos River (drainage basin 38,300 square miles), and Rio Salado. The Rio Chama contributes more discharge than the Rio Puerco, which has an 18,900-km^2 drainage basin area. Today, much more sediment is delivered to the Rio Grande by the Rio Puerco and the Rio Salado than by the Rio Chama. This is because there are dams on the Rio Chama. In contrast, the Jemez River, which has a drainage area of 2690 km^2 and a discharge that is almost twice as large as the Rio Puerco, contributes little sediment to the Rio Grande because of the presence of the Jemez Canyon Dam (Bullard and Wells, 1992).

The concentrations of sediment in the ephemeral Rio Puerco have exceeded 680,000 mg L^{-1}. Seventy-five percent of this sediment is sand (Nordin, 1963), which represents about 68% of the sediment by weight. In comparison, turbulent water floods are 1 to 40% sediment by weight, and the debris flow is much higher, being 70 to 90% sediment by weight (Costa, 1984). The highest daily suspended sediment load for the Rio Puerco is 2,032,000 tons. The Rio Puerco lacks discharge of water or sediment for many days of the year. It is interesting to note that from the Rio Puerco, perhaps 82% of the sediment transport takes place during storm events that recur about once a year (Leopold et al., 1964). In 1987, the maximum daily suspended sediment load was 212,000 tons (Bullard and Wells, 1992).

Fluvial Responses to Changing Sediment and Water Discharge.   In the natural landscape, the fluvial system is a dynamic geomorphic component. The river attempts to adjust itself to changes in the fluvial system, such as stream slope, sinuosity, and water and sediment discharges (Schumm, 1977). The stream tries to maintain a condition of dynamic equilibrium, or energy balance, whereby the discharge of a river is balanced with the discharge of the sediment load. A stream is capable of transporting a particular volume of sediment for any given discharge. A decrease in availability of sediments may occur if they are trapped in reservoirs. Thus the channel will tend to erode, and incising of the banks and the bed may occur. Increase in the sediment load with a decrease in flow will cause the river to respond by deposition of sediment load, aggradation, reduction in channel slope, and changes in stream patterns (e.g., meandering and braiding). Incisions and additional changes in gradient will eventually follow, which will result in changes in channel pattern as the river strives to attain a condition of dynamic equilibrium. Many things may happen as a result of these altered river conditions, such as loss of water from the channel, change in the level of shallow groundwater, change in direction of groundwater flow, and even impacts on riparian vegetation communities. Thus, stream flow regulated by dams causes several physical components of a fluvial system to adjust to new sediment and water discharge conditions including sediment delivered by tributaries (Bullard and Wells, 1992).

Factors Affecting Pattern of Flow.   High sediment discharges from tributaries may cause problems such as frequent changes in channel pattern from straight or meandering to braided, channel capacity may decrease, and the floodplain area may aggrade and increase. In addition, the levels of groundwater may rise and stands of phreatophytes may be established. These transpire large quantities of water. As a result, dramatic losses in water delivery to downstream reservoirs in the Rio Grande may occur. Federal agencies attempted to counteract these losses by maintaining a well-defined channel for the Rio Grande (Bullard and Wells, 1992).

Influence of Tectonics on River Behavior. The fluvial system may be influenced dramatically by tectonics. For example, tectonic activity such as uplifts and warping may cause changes in channel pattern, sediment transport characteristics, and channel gradient. Overall, the geomorphic evolution of a fluvial system may be affected. For example, the Rio Grande Rift was so affected. Ouchi (1983, 1985) found that in the natural channel of Rio Grande there were measurable changes due to active tectonic uplift near Socorro. These included changes in gradient, sinuosity, and sediment transport characteristics. He found that tectonic disturbance and resultant changes in flow characteristics and sediment transport ability of the river caused the channel gradient to decrease upstream. There were also changes downstream due to tectonic disturbances. From these studies it is evident that tectonic activity within the Rio Grande Rift may change the behavior and pattern of the Rio Grande through time independent of changes that might be caused by human activity (Bullard and Wells, 1992).

Effects of Dams: Changes in Sediment Load and Deposition. Two main sources of sediments available for fluvial transport after the closure of Elephant Butte Dam were identified by studies of Flock (1931) and the International Boundary and Water Commission (1932). These were the river bed, subject to scour by the clear waters released by the dam, and sediment derived from the tributaries. During the first 15 years after closure of the dam, the river scour in the Rio Grande was about 0.6 m in depth. It extended about 160 km downstream from Elephant Butte through the Mesilla Dam reach (Bullard and Wells, 1992).

Small deltas were formed by sediment entering the Rio Grande from the tributaries. These deltas resulted in localized channel migrations or shifting. This was caused by the low magnitude and short duration of reservoir releases. These were incapable of effectively transporting the sediments. Within the canyon the small tributary deltas partially dammed the reach and thus formed a local base-level condition which results in backwaters above, and rapids over, the tributary deltas. The channel bottom was armored by gravel in some reaches where the Rio Grande was capable of transporting only the sand fraction supplied by tributaries. The gravel fraction remained behind as an armor on the channel bed (Bullard and Wells, 1992).

The Rio Grande from Cochiti to San Felipe consisted mainly of braided channels separated by bars and islands before the Cochiti Dam construction. These braided channels were composed of coarse gravel and cobbles. The active riverbed was sand at low discharges and sand and gravel at higher flows. The Rio Grande was a sandbed stream below the confluence of the Rio Grande with the Jemez River. The rate of flow that could balance sediment input and effective sediment transport was in the range of 6000 ft^3/s for the pre-Cochiti River (Bullard and Wells, 1992).

Effects of Dams on Flow. The physical nature of the Rio Grande channel is affected profoundly by construction of the Cochiti Dam, which altered the sediment load characteristics (Lagasse, 1980, 1981). Initially, the channel degraded until an armor of gravel was established on the channel bottom. This locked the channel in position and created stable conditions immediately downstream from the dam. Regulated discharge from Cochiti Dam was insufficient farther downstream to transport sediment supplied to the main stem by small, unregulated tributaries. This resulted in an unstable channel configuration. It was evident that small unregulated tributaries could influence the behavior of the Rio Grande with respect to channel platform, channel cross section, sediment transport, aggradation, and stability (Bullard and Wells, 1992).

The reduction in Rio Grande flow below the channel-forming discharge (6000 ft^3/s) enabled small tributaries to affect the Rio Grande channel. The average discharge at San Felipe before the closure of Cochiti Dam was 1374 ft^3/s 24 km downstream from the Cochiti Dam. Since closure the average discharge has been greater, 1626 ft^3/s. The apparent increase in average downstream discharge through the San Felipe gauge since dam closure may be misleading. The implication of this difference in discharge is not immediately obvious. There has been more consistency in flow of the Rio Grande since the closure of Cochiti Dam, and there has been a reduction in the frequency of large, seasonal floods. Since 1979, 8 of 10 years have produced runoff well above the 20- to 25-year average, with corresponding increases in discharge. Long-term cycles in flow have been damped by dam construction (due to climatic fluctuations), although the cyclicity has not been eliminated entirely (Bullard and Wells, 1992).

The emplacement of the Cochiti Dam has eliminated channel-forming discharges and flooding. This has implications for sediment transport and river behavior. The absence of seasonal floodwaters that provided necessary conditions for germination, establishment, and maintenance of many riparian species may have been affected (Bullard and Wells, 1992). Since (in 1992), the Cochiti Dam had been in existence for less than 20 years, Bullard and Wells state that the long-term effects on the riparian vegetation and associated faunal communities may not yet have been apparent.

Effects of the Dam on the Rio Grande Upstream from Elephant Butte Reservoir. The reaches downstream from a dam are not the only reaches affected by dams and reservoirs. Upstream changes are first, that the reservoir produces a new base level for upstream reaches of the river. Second, the upstream rise of the base level of the reservoir decreases the slope of the upstream reach and thus changes the fluvial geomorphology and hydraulic geometry (Bullard and Wells, 1992).

Two major effects resulting from the creation of Elephant Butte Reservoir are siltation and loss of water to the channel. The Rio Grande formed a delta near San Marcial. Artifical base-level conditions created by the reservoir caused a reduced river slope upstream, and this was accompanied by a decrease in flow velocity and sediment transport. As a result, there was deposition and formation of a large, fine-grained delta. This caused the development of a system of anastomosing channels and the loss of a single, distinct channel. As a result, there were encroachment and establishment of thick stands of riparian vegetation (Bullard and Wells, 1992).

Decrease in water delivered to Elephant Butte was caused by losses of water and riparian vegetation and infiltration into the channel of deltaic sediments (Bullard and

Wells, 1992). The effects of the rectification program on floral and faunal communities outside the designated floodway are unknown. On rare occasions the river waters extended outside the levee system. For plant species that depended on seasonal flooding for germination and establishment, the implications are obvious. However, specific effects are still not known (Bullard and Wells, 1992).

Materials transported through the Rio Grande system are being deposited in the Elephant Butte Reservoir. Examples of these are cadmium, mercury, lead, uranium, and pesticides. These substances are bound primarily to sediments. However, an unknown amount is being cycled from the sediments to the water column in the reservoir (Bullard and Wells, 1992).

Diversions of Flow

New Mexico. A large portion of the flow from the Elephant Butte Dam is diverted to irrigate croplands in New Mexico. The remainder and return flow then reaches El Paso at an annual rate of 547,000,000 m^3. As the flow reaches the American Diversion Dam, 332,000,000 m^3 has been diverted annually to the American Canal, which is the main supply canal for the El Paso valley.

Mexico. The diversion in Mexico has amounted to 65,000,000 m^3 annually, which is used to irrigate the Juarez valley along with the shallow groundwater and municipal sewage. After diversion, the flow of the Rio Grande is reduced to 155,000,000 m^3 annually. The flow gradually increases due to the collection of return flows and municipal sewage water discharge from several plants from El Paso and adjacent communities. The sewage water from Ciudad Juarez is discharged into irrigation canals and, to a limited extent, into drainage ditches but not directly into the Rio Grande. When the flow reaches Fort Quitman, storm runoff from small creeks is added to the flow of the Rio Grande. The Rio Conchos that originates from Mapimi drainage basin from the state of Chihuahua carries an average annual flow of 909,000,000 m^3 at a point of inflow into the Rio Grande near Ojinaga, Mexico. This flow is slightly greater than the annual release from Elephant Butte Dam and forms the main flow of the Rio Grande in the reach between Presidio and Amistad Dam.

Alterations in Flow

The Rio Grande waters reaching Texas (alloted under the terms of the 1938 Rio Grande Compact) are provided by releases from Elephant Butte Reservoir, 125 mi upstream from El Paso in New Mexico. Before the waters of the Rio Grande reach the cities of El Paso and Juarez, most of the flow is diverted for irrigation and municipal uses at the American Canal in Texas and the Acequia Madre Canal, the main irrigation canal in Juarez, Mexico. Downstream of El Paso, most of the flow consists of occasional storm runoff, treated municipal wastewater from El Paso, and irrigation return flows. From El Paso/Ciudad Juarez to Presidio/Ojinaga, the flow is intermittent. Inflow from the Rio Conchos near Presidio/Ojinaga provides over three-fourths of the flow to segment 2306—the Big Bend of the Rio Grande (TNRCC, 1994).

In the middle reach of the Texas part of the basin, construction of the International Falcon Dam in 1953 and International Amistad Dam in 1968 resulted in the most significant hydrologic modifications to the flow of the Rio Grande/Rio Bravo. Both of these dams were constructed by the International Boundary and Water Commission (IBWC) under the terms of the 1944 Water Treaty between the United States and Mexico (TNRCC, 1994).

In the lower Rio Grande valley the floodplain of the Rio Grande narrows considerably and is less than a mile wide in northwestern Starr County, where the river leaves the impoundment of Falcon Dam. The floodplain is about 6 mi wide in Hidalgo County but then broadens into a wide delta fronting the Gulf of Mexico near the mouth of the river in Cameron County. The delta includes old meanders of the river that are fringed by natural levees (Judd, 1994). The river discharges directly into the Gulf of Mexico, except during floods, when much of the water is diverted into flood channels and thence to the Laguna Madre (TNRCC, 1994). However, at the present time there is little or no flow into the Gulf of Mexico, due to diversion for various uses.

CHEMICAL AND PHYSICAL CHARACTERISTICS OF THE WATER

Chemical and Physical Characteristics of the Rio Grande at El Paso, Texas (1987–1988)

Location: latitude 31° 48′ 10″, longitude 106° 32′ 25″, El Paso County, Hydrologic Unit 13030102, at a gauging station on the downstream side of the Courchesne Bridge, 5.6 mi upstream from the Santa Fe Street-Juarez Avenue bridge between El Paso, Texas and Ciudad Juarez, Mexico, and 1.7 mi upstream from the American Dam (Buckner et al., 1989). See Table 3.1.

The stream flow in 1988 varied from 137 ft^3/s in February to 1750 ft^3/s on June 16. The specific conductivity varied from 909 µS cm^{-1} on August 19 to 1870 µS cm^{-1} on February 17. pH varied from 7.5 on August 19 to 8.10 on June 16. The temperature varied from 10.0°C on January 29 to 21.0°C on June 16. Total hardness varied from 220 ppm on August 19, 1988 to 410 ppm on October 21, 1987. Dissolved calcium varied from 63 mg L^{-1} on August 19, 1988 to 120 mg L^{-1} on October 21 and November 16, 1987. Dissolved magnesium varied from 15 mg L^{-1} on June 16 and August 19 to 27 mg L^{-1} on January 29 and February 17. Dissolved sodium varied from 100 mg L^{-1} on August 19 to 260 mg L^{-1} in January and February. Potassium dissolved varied from 5.9 mg L^{-1} in August to 11 mg L^{-1} on January 29. Alkalinity of the water varied from 131 mg L^{-1} on September 22 to 246 mg L^{-1} in October and November 1987. Dissolved sulfate varied from 170 mg L^{-1} in August to 420 mg L^{-1} in February 1988 and December 1987. Chlorides varied from 77 mg L^{-1} in August to 190 mg L^{-1} on October 21, 1987. Silica varied from 13 mg L^{-1} in May and June to 24 mg L^{-1} in October and December 1987 (Buckner et al., 1989).

Toxic Substances in the Rio Grande

Metals and other toxic materials have accumulated in the Rio Grande sediments and in some cases in fish. This has been due to mining activities in the headwaters and mainly to agricultural activities and associated industries in the remainder of the river basin.

TABLE 3.1. Water Quality Data, Water Year October 1987–September 1988, Rio Grande Basin at El Paso

Date	Time	Instantaneous Stream Flow (ft³/s)	Specific Conductance (μS cm⁻¹)	pH (standard units)	Water Temperature (°C)	Total Hardness (mg L⁻¹ as CaCO₃)	Hardness of Noncarbonated Water, Total Field (mg L⁻¹ as CaCO₃)	Dissolved Calcium (mg L⁻¹ as CaCO₃)	Dissolved Magnesium (mg L⁻¹ as Mg)
Oct. 21	10:45	326	1760	7.80	14.5	410	160	120	26
Nov. 16	8:30	276	1720	8.00	11.0	400	160	120	25
Dec. 15	9:30	210	1800	7.90	—	380	140	110	26
Jan. 29	13:00	230	1810	7.90	10.0	390	150	110	27
Feb. 17	8:45	137	1870	8.00	11.0	390	140	110	27
Apr. 21	7:00	890	1030	7.80	16.5	240	70	70	16
May 19	7:20	480	1090	8.00	11.0	250	67	75	16
June 16	7:10	1750	949	8.10	21.0	230	58	66	15
July 21	7:10	300	1010	7.80	20.0	240	65	69	16
Aug. 19	7:35	1200	909	7.50	20.0	220	70	63	15
Sept. 22	7:15	812	1120	7.90	15.5	250	120	72	17

Date	Time	Dissolved Sodium (mg L⁻¹ as Na)	Sodium Adsorption Ratio	Dissolved Potassium (mg L⁻¹ as K)	Alkalinity of Water Whole Field (mg L⁻¹ as CaCO₃)	Dissolved Sulfate (mg L⁻¹ as SO₄)	Dissolved Chloride (mg L⁻¹ as Cl)	Dissolved Silica (mg L⁻¹ as SIO₂)	Dissolved Solids, Sum of Constituents (mg L⁻¹)
Oct. 21	10:45	240	5	9.8	246	380	190	24	1140
Nov. 16	8:30	220	5	9.4	246	380	170	23	1090
Dec. 15	9:30	240	6	9.0	239	420	200	24	1170
Jan. 29	13:00	260	6	11	239	410	200	20	1180
Feb. 17	8:45	260	6	10	245	420	170	20	1160
Apr. 21	7:00	120	3	6.9	171	210	95	15	635
May 19	7:20	130	4	6.6	187	220	100	13	673
June 16	7:10	110	3	6.2	169	190	78	13	580
July 21	7:10	120	4	6.5	174	210	90	15	631
Aug. 19	7:35	100	3	5.9	149	170	77	17	537
Sept. 22	7:15	130	4	6.9	131	230	110	18	662

Source: Modified from Buckner et al. (1988).

121

Metals have been found in water samples from the headwaters of Falcon Lake downstream to Laredo–Nuevo Laredo and at the Piedras Negras discharge. Barium and iron were above detectable levels at all three sites. Manganese was also found in water samples. Except for manganese from the Piedras Negras discharge, levels detected are less than the historical averages for samples from the Rio Grande from the Gulf of Mexico to the confluence of the Rio Conchos near Presidio (Buzan, 1990).

At the headwaters of Falcon Lake and at Piedras Negras, the discharges contained arsenic, barium, chromium, copper, lead, manganese, mercury, nickel, and zinc, which were at levels above the limits of detection. Concentrations of metals other than manganese were less than the historical averages. In the 12 sediment samples analyzed for metals in the Rio Grande downstream from Presidio, only cadmium was below the level of detection. The concentrations of nine of these metals exceeded the amounts from the headwaters of Falcon Lake. The concentrations of lead (93 mg kg^{-1}) and zinc (160 mg kg^{-1}) were over twice as great as values from Falcon Lake headwaters and the historical averages for the Rio Grande downstream from Presidio. Metals-in-sediment data from the Piedras Negras discharge suggest the possibility of periodic wastewater discharges with metals from one or more industrial sources (Buzan, 1990).

Pesticides-in-water were sampled at the same locations: at the headwaters of Falcon Lake, downstream from Laredo, and at the Piedras Negras discharge. None of the organic compounds tested were above detection levels at the three sites. Pesticides-in-sediment were sampled at the headwaters of Falcon Lake and at the Piedras Negras discharge. Of the organic compounds that were tested none were in the headwaters of Falcon Lake. However, chlordane in both the cis isomer (7.0 µg kg^{-1}) and the trans isomer (5.0 µg kg^{-1}), and p,p'-DDE (11 µg kg^{-1}) were detected in the Piedras Negras discharge sediment (Buzan, 1990).

Heavy metal contamination in the Rio Grande is quite significant. The El Paso region has very significant accumulation of dissolved chromium and dissolved lead, probably associated with colloids. Beginning with segment 2314, which stretches from the International Dam in El Paso County to the New Mexico state line in El Paso County, in November 1992 the Texas Water Commission (TWC) reported the following numbers (Briones, 1994):

Segment 2314—Metals in Number of River Sediment Samples

River sediment metals	8
Pesticides	8
Fish tissue metals	3
Pesticides	3

For segment 2308, which stretches from Riverside Diversion Dam in El Paso County to International Dam in El Paso County, the report quotes the following number of samples comprising the database. A total of 19 samples were taken for river-water metals. Seven of the samples produced metals. Other toxicants found in the river water were pesticides (seven samples). The data available are not very complete for these contaminants (Briones, 1994).

Chromium. Forty-two samples of water were collected at Courchesne Bridge (1.7 mi from American Dam in El Paso) by the Texas Natural Resource Conservation Commission (TNRCC) for dissolved chromium. Of the 42 collected, 13 were within limits, 29 were out of limits (TNRCC, 1994). The highest reading was November 3, 1981, which is a reading of concern according to the human health criteria in water and fish. In data from May 12, 1987, lead concentrations were almost two-fold greater than the Criteria in Water for Specific Toxic Materials.

There is not much concern for chromium in mud and total chromium in water. The concentrations are close to the TWC criterion limits in water for specific toxic chemicals relating to human health protection and to criterion limits in water for specific toxic materials relating to aquatic life protection. These amounts pose no immediate threat to aquatic life but could pose serious ecological problems if they accumulate over time (Briones, 1994).

Five samples of total chromium were obtained from November 3, 1981 to December 14, 1983 (TNRCC, 1994). Two of these samples were within limits and three were out of limits. There were seven mud samples collected between September 6, 1978 and June 16, 1987. All of these samples were within limits of the Criteria in Water for Specific Toxic Materials relating to Aquatic Life and Human Health Protection (Briones, 1994; TNRCC, 1994; TWC, 1992b).

Dissolved chromium was of concern from Brownsville to Laredo as all of the dissolved chromium readings were in excess of limits (TNRCC, 1994). Of 81 readings taken in the El Paso region, 51 readings are out of limits (TNRCC, 1994). The major concern is Courchesne Bridge. Dissolved chromium accumulation doubled from 1981 to 1991 in the Rio Grande from Brownsville to El Paso.

Lead. Thirty-nine samples of dissolved lead were collected in the Fort Quitman area (1.5 mi downstream from Old Fort Quitman at the gauging station on the rectified channel of the Rio Grande (TRNCC, 1994). These samples were collected between November 6, 1981 and September 11, 1991. Of 39 taken, 11 samples were within the limits for Aquatic Life Protection and Human Health Protection. Of the 39 samples, 28 were outside the limits set for the Human Health Protection Criterion. The violation exceeded the criterion as much as 1 μg L^{-1} (Briones, 1994). For readings from November 6, 1981 to September 11, 1991, three of the five readings per year, the coefficient of variation was 68.113% (Briones, 1994).

Lead exceeded the screening criteria in water (TWC, 1992a) in segment 2306 (the Big Bend) from the confluence of the Rio Conchos with the Rio Grande and into Val Verde County. In Table 3.2 the database for this segment is given (Briones, 1994). Sixty samples were reported in the TNRCC report in 1994. Forty-three samples exceeded the screening criteria in water. These samples were collected between October 7, 1981 and August 21, 1991, with the average mean being 2.846 μg L^{-1} for dissolved lead at Foster's Ranch west of Langtry. The average mean is within the range, probably because 17 of the samples collected were <5 μg L^{-1}. Thirty-nine of the 60 samples were collected from Foster's Ranch inside Val Verde County. The mean was 2.846 μg L^{-1}. The highest accumulation of lead, which violates the Marine Chronic Criteria for Aquatic Life Protection, was at Foster's Ranch near Langtry.

Between October 7, 1981 and August 21, 1991, 39 samples were collected at Foster's Ranch. Twelve of the 39 samples were within limits and 27 samples violated the

TABLE 3.2.   *Segment 2306—Metals
in Number of River Sediment Samples*

River-water total metals	2
Dissolved metals	39
Pesticides	18
Volatile organics	0
Acid extractables	0
Base neutral extractables	0
Ambient toxicity testing	2
River-sediment metals	11
Pesticides	22
Fish tissue metals	6
Pesticides	24

Source: TWC (1992a) in Briones (1994).

Human Health Criterion for Water (TNRCC, 1994). An analysis of these data indicates that there was a mean 2.846 μg L⁻¹, a standard deviation of 2.337 μg L⁻¹, and a high of 10 μg L⁻¹ on January 31, 1990 (Briones, 1994).

One of the major readings for lead concentration that was slightly above the human health protection criterion was at the Rio Conchos confluence with the Rio Grande in Presidio County. This was taken September 12, 1979. Another sample taken on July 23, 1994 at the same location was 10 μg L⁻¹ which is above the human health protection criterion. Lead in sediments 2.4 mi upstream from the Rio Conchos confluence was found to be high. A sudden drop from September 12, 1979 to September 23, 1980 was found in the actual readings. Then readings were variable for September 1980 to May 22, 1989 (Briones, 1994). High levels of lead in fish tissue samples were also found in this segment. Data on fish tissues for the Rio Conchos confluence is from September 13, 1988, and for the tidal shore in Brownsville it is from 1984 and 1988 (Briones, 1994).

In the Laredo–Brownsville region, the TNRCC Station 13698, 1.1 mi downstream from the highway bridge between Laredo and Nuevo Laredo, had 19 samples of dissolved lead collected between October 21, 1981 and August 27, 1986. Of these 19 samples, four were within limits, with a mean of 1.421 μg L⁻¹ and a standard deviation of 1.042 μg L⁻¹. According to the TWC November 1992 report on the Rio Grande basin, total dissolved solids were 1000 μg L⁻¹. Fecal coliforms were 200 per 100 mL of water for segment 2304, which extends from the confluence of Arroyo Salado from Mexico in Zapata County to Amistad Dam in Val Verde County (Bowles, 1983; Buzan, 1990; TWC, 1992a).

Slightly above the criterion of 5 mg kg⁻¹ were lead readings from September 2, 1981 to May 22, 1989 in the Rio Conchos confluence. A high reading of 36.4 mg kg⁻¹ occurred on September 12, 1979 at Rio Conchos confluence. A flood that reached 50 ft above normal on November 5, 1978 was probably the cause (Big Bend Natural History Association, 1993; Briones, 1994). Between January 12, 1978 and May 10, 1988, there were 15 readings of 50 μg L⁻¹ of total lead at Falcon Dam. On June 24,

1986 10 µg L^{-1} was recorded, which was the lowest data point. Three readings above the human health criterion occurred at Foster's Ranch on October 12, 1983; November 2, 1988; and January 31, 1990 (Briones, 1994; TNRCC, 1994). Dissolved lead remained the same for a 10-year period at Brownsville, stabilized at Laredo, and increased by 5 µg L^{-1} at Presidio. Presidio has been cited as an area of concern for lead accumulation by the International Boundary and Water Commission Report released October 19, 1994 (Briones, 1994; McLemore, 1994).

Lead was found to be of slight to moderate potential for toxicity at the Rio Conchos confluence, Foster's Ranch, and Falcon Dam. Sites at the Courchesne Bridge in El Paso, Laredo Nuevo–Laredo Bridge, and El Jardin Pump Station in Brownsville are also areas of concern for dissolved chromium.

From Laredo to Falcon Dam and Brownsville, lead continued to accumulate. High readings of 50 µg L^{-1} were found at Falcon Dam from January 12, 1978 to May 10, 1988. These amounts were out of limits, as they had a mean of 48 and a standard deviation of 10. At El Jardin Pump Station in Brownsville, the averages for dissolved lead are 5 µg L^{-1} or below. This is close to the criterion for human health protection. It should be pointed out that on February 14, 1990 the reading was 20 µg L^{-1}. This is four times the human health protection criterion (Briones, 1994).

In contrast, dissolved chromium was below the criterion of 50 µg L^{-1} as the results for four readings were 10 µg L^{-1} each at the El Jardin Pump Station, Brownsville, with other points below 5 µg L^{-1}. At Brownsville, tidal shore samples had the highest reading of dissolved chromium. The highest reading of total chromium, 30 µg L^{-1}, occurred on August 18, 1982 at Laredo–Nuevo Laredo. Chromium is also under the human health criterion (Briones, 1994).

EFFECTS OF DIVERSION, HUMANMADE CHANGES, AND SOURCES OF POLLUTION

Diversion and Return Flow from Irrigation

The Rio Grande can be thought of as four different rivers, each with its own set of problems and characteristics (Hodge, 1994). The first river is the Rio Grande as it begins at its headwaters in the San Juan Mountains of central Colorado (approximately 12,000 ft elevation). It is a mountainous stream and is the home of numerous wildlife species. Near Silverton, Colorado the river spills into the broad San Luis Valley. As early as 1990, during peak growing seasons, farmers were draining this part of the river dry. The 200 mi between Española, New Mexico, north of Santa Fe, and Elephant Butte Reservoir is used for irrigation. The Rio Grande is a desert river through the first section of the river. This section is 635 mi with decreasing latitude and elevation (Van der Leeden et al., 1990). The amount of water being discharged in this part of the river per square mile of basin is considered to be one of the least in a list of the world's principal rivers. The discharge is a mere 0.02 ft^3/s, whereas in the Mississippi–Missouri system it is 0.52 ft^3/s (American Rivers, 1993; Briones, 1994).

From Elephant Butte to Fort Quitman is the second section of the Rio Grande. In the 110 mi from Elephant Butte to the Texas state line, three dams divert the river into ditches that furnish irrigation water for more than 90,000 acres (Briones, 1994).

The third Rio Grande extends from Fort Quitman to Lake Amistad near the city of Del Rio. From Fort Quitman to the confluence of the Rio Conchos, the first 200 mi is usually baked dry in the desert sun from October through January when irrigation water is turned off. Just above Presidio, the Rio Conchos injects new life into the dying river (TWC, 1992a). For all practical purposes, the American river known as the Rio Grande ends at Presidio. From this point on, about two-thirds of the water that enters the Rio Grande originates in Mexico in the Sierra Madre, Occidental of Chihuahua, and Durango. A more appropriate name for this reach would be El Rio Bravo del Norte because the Rio Conchos confluence brings water from the Sierra Madre, Occidental of Chihuahua, and Durango. The cotton that grows along the banks of the canal has been sprayed with DDT from the American side. From the Presidio Mine of the American Metal Company of Texas, located at Shafter, Texas, copper, lead, and silver were mined in 1941 (personal records, Cosmet Briones, 1941–1954; Briones, 1994).

Tributaries of the Rio Grande: Cibolo Creek, Rio Conchos
Two miles above Presidio, Cibolo Creek, which runs through the edge of the mine, empties 21 mi downstream into the Rio Grande. The Rio Conchos, which is fed by five tributaries, enters the Rio Grande. Late summer tropical rains and spring snowmelt result in peak flows. The Big Bend region of Texas and Mexico furnishes water for recreational purposes for river rafters. It also is an important source of irrigation water for the lower Rio Grande valley. The Rio Grande borders Big Bend National Park for 107 mi. The 196 mi of river designated "wild and scenic" in 1978 is managed by the National Park Service (Hodge, 1994; Maxwell, 1990).

Effects of Terlingua Creek and Pecos and Devil's Rivers
Terlingua Creek and Pecos and Devil's rivers are among the largest rivers that enter the Rio Grande between the Rio Conchos and Lake Amistad at Del Rio. The International Boundary and Water Commission and La Commission Internacional de Limites y Aguas read the gauging stations on both sides of the border weekly (Briones, 1994).

Contributions of Flow from U.S. and Mexican Tributaries
Thirty percent of the Rio Conchos flow and water coming from the tributaries are credited to the United States. Similarly, Mexico is credited with water coming from tributaries on its side and with some water coming out of the United States. The water in Lake Amistad at any one time is divided between two countries based on where the inflow originates (Briones, 1994).

Pollution of the "Fourth Rio Grande"
The fourth Rio Grande (lower section) has become a "conduit for disease" by the time it winds through the heavily populated country and flows into the Gulf of Mexico (*Dallas Morning News*, August 9, 1994). In many segments the river has been greatly changed, as it has been diverted, ditched, and channelized. As a result, the ri-

parian habitats are lost or altered so that they no longer resemble a natural river (American Rivers, 1993). Furthermore, the water has been overappropriated and contaminated with industrial wastes, agricultural pesticides, and runoff. Associated with the sediments of the river have been found municipal sewage, heavy metals, mine wastes, including cyanide and acidic runoff, as well as plutonium and other nuclear wastes from the labs at Los Alamos (American Rivers, 1993; TWC, 1992a). Mexican factories (*maquiladoras*) located on the Mexican side of the Rio Grande at Juarez and Matamoros serve the textile, aircraft, and refractory industries. On the U.S. side at El Paso, Del Rio, Laredo, and Brownsville are some facilities involved in these types of industries. Some malquiladoras have been set up by the state of Chihuahua and by American corporations along the Rio Conchos. The manufacturing of paints is the speciality of these malquiladoras, many of which have been found to be dumping their effluent into the Rio Conchos and have been cited for the dumping (American Rivers, 1993; Briones, 1994).

CAUSES OF RIVER DETERIORATION

Animal and Human Waste

The reach of the river studied in the upper Rio Grande extended from Caballo Dam in New Mexico, downstream approximately 100 river miles (171 km) to the American Diversion Dam at El Paso, Texas. In this area when the study was made there were more than 90 discrete sources of water pollution. Fifty-two are domestic sewage treatment facilities. Only three of these, the sewage treatment plants at Hatch and Las Cruces, New Mexico and at Anthony, Texas, discharge to the surface waters and can be considered point sources of pollution. There are two sources of agricultural pollution: irrigation drains, of which there are nine that are major, and animal confinements. Only one of the drains is a true point source to the Rio Grande in this region.

Although the animal confinement facilities are not permitted to discharge wastes to the surface waters, there are many accidental sources of pollution during floods, such as illegal waste disposal in irrigation laterals and arroyos and faulty operations of spray irrigation systems.

Nearly all the industrial sources are also related to agriculture. Most are chili-, cotton-, or pecan-processing operations which operate seasonally. Examples of major industrial point sources are an El Paso electrical company steam electrical station and the El Paso natural gas coolant plant. There are also three municipal sewage treatment plant discharges.

One sewage treatment plant served a population of approximately 45,400 in 1972. It also received wastes from a number of food-processing operations. It had primary and secondary clarifiers, digesters, and trickling filters, which removed biological oxygen demand (BOD) and chemical oxygen demand (COD) to 40 and 50%, respectively. Unfortunately, the wastewater flow through the plant was almost entirely anaerobic and thus limited the amount of BOD reduction that could be achieved. Shock loading originating from food-processing facilities constituted an additional operating problem. During the period of study, effluent varied from 31 ppm in September 1972 to 144 ppm in December 1972. The pH was always circumneutral.

Overgrazing

After the Civil War, western migration of Anglo-European settlers greatly increased the population of the Rio Grande valley. With the immigrants came an intensification of both ranching and farming activities. By 1880, nearly all land in the Rio Grande valley that could be irrigated was developed and water shortages were reported (Sodrensen and Linford, 1967). By the end of the nineteenth century, grasslands were seriously overgrazed, and stream flows had become more erratic, soil erosion was accelerating, and the stream channels were wider and shallower. River flow in the Mesilla valley is known to have ceased at times during 1879, 1891, 1894, and 1896 (Baldwin, 1938). This dewatering began 37 years before the construction of Elephant Butte Reservoir (completed in 1916), the first large reservoir built on the Rio Grande. Lee (1907) described the Rio Grande north of El Paso as "mainly a floodwater stream subject to great fluctuations in volume" (Sublette et al., 1990).

The effects of overgrazing in the Rio Grande drainage are clearly depicted in the Rio Puerco valley, where seeps and occasional storm waters flow through an 8.5-m-deep vertically walled arroyo. Prior to 1885, this deeply incised arroyo did not exist. The period of its cutting coincides with the maximum grazing of livestock in the valley (Bryan, 1928). The increased sediment load of the Rio Puerco was, and continues to be, deposited in the Rio Grande within a section of the river that is naturally aggrading. This in turn has increased the incidence of flooding and requires numerous elaborate water projects to compensate (Sublette et al., 1990).

Dams

During the twentieth century, numerous large dams have been constructed in New Mexico on the Rio Grande and its major tributaries. These include Caballo, Elephant Butte, Bluewater, Jemez Canyon, Cochiti, Galisteo, McClure (in Santa Fe), Abiquiu, El Vado, and Heron. In addition, numerous smaller dams and irrigation diversions and sediment control dams have been constructed. Water releases from these reservoirs fluctuate greatly and downstream flows are often critically depleted (Sublette et al., 1990).

Changes in Stream Flow—Irrigation, Droughts

It is evident that stream flow in the study area (Caballo Dam to el Paso) has declined over time. Although variation in monthly (or yearly) flow exists, when streamflow data are averaged over "water-decades" the trends become more apparent (Edwards and Contreras-Balderas, 1991). Extensive irrigation (especially since 1940) has had the effect of lessening the flow in the river. Irrigation removes water from the river's flow. The water flows instead into the floodways and irrigation systems and eventually enters one of the Laguna Madres of Texas or Mexico to the northeast or southeast of the river's mouth or merely evaporates into the atmosphere. The extent of the change in water flow over time is easily seen by comparing the flow at Laredo (above the heavily irrigated area) with the flow at Brownsville and Matamoros. Another factor that contributes radically to decreasing flow is severe drought such as that of the 1950s. Although the 1970s water decade was one of the most "water-rich" decades of the century, the yearly average flow of the lower part of the Rio Grande was only one-third to one-half of what it had been in former years (Edwards and Contreras-Balderas, 1991).

Exotic Fish Species
Proliferation of exotic species may have altered natural associations of fish in many areas, particularly in the Rio Grande (Edwards and Contreras-Balderas, 1991). Large changes have occurred in the structure of fish populations in the Rio Grande system. Because of the major environmental changes that have taken place—in reservoirs, water removal, introduction of exotic species, and pollution—major faunal changes would be expected. These changes have occurred in the lower Rio Grande (Edwards and Contreras-Balderas, 1991). In the upper and middle reaches near the Edwards escarpment, most of the Waco area sites were located where impoundments have been created during the past 30 years. There has been a statewide (Texas) trend in reduction of lotic-adapted taxa with narrow habitats. These are species such as darters, minnows, suckers, and catfish. Opportunistic species (mosquitofish and silversides) have increased (Anderson et al., 1995).

Agricultural Chemicals and Petrochemicals
Since 1950 there has been an increased impact in the lower Rio Grande region by agricultural chemicals and petrochemicals. A number of chemicals which can be found in quantities approaching or exceeding the guidelines of the U.S. Environmental Protection Agency for the safety of fishes and other aquatic organisms have been found in the lower stretches of the river. The changes in faunal abundance due to the presence of these chemicals alone have not been quantified, but the effect on the indigenous fauna seems to be in concert with the observed changes in stream flow. Evidence of accumulation of toxic chemicals by some fish of the lower Rio Grande has been presented by White et al. (1983) and Andreason (1985) (Edwards and Contreras-Balderas, 1991).

Human Cultural Activities and Water Use
Maximizing the use of water for human cultural activities has changed the fish fauna of southwestern United States and northern Mexico and possibly caused to be lost forever components of the native fish populations.
    Tamaulipan Biotic Province. The lower Rio Grande in Tamaulipan Biotic Province is similar to other southwestern U.S. and northern Mexican streams as to the historical sequences involving their fish faunas. At present there is a decreasing stream flow in streams and rivers in increasingly mesic regions. Formerly, this was never a problem of significant magnitude. Eventually these streams will probably become streams with fewer fish in less diverse assemblages (Edwards and Contreras-Balderas, 1991).

AQUATIC LIFE: ELEPHANT BUTTE DAM TO CABALLO DAM

Various studies have been made over time of the aquatic life in the Rio Grande. The ecosystem has been studied in greatest detail between the Elephant Butte Dam and Caballo Lake. The aquatic life in this area gives us the best representation of the functioning ecosystem in the Rio Grande. The Rio Grande in this section varied in length from 20 to 25 km depending on the elevation of Caballo Lake. The Rio Grande is an eighth-order river (Bristow, 1993) in this area.

The upper study site ran from Elephant Butte Dam to Highway 51 (Figure 3.3). This site is characterized by the largest substrate particle size in the entire tailwater and by the occurrence of bedrock substrate. The pool habitat is a stilling basin below the dam in the upper study area. The pool/riffle ratio area in this reach is the lowest of the three study areas, being 0.13:1.00. The highest stream gradient is found in this area. It is approximately 1.18% (Bristow, 1993).

In the reach of the Rio Grande between Elephant Butte Dam and Caballo Lake, the primary production was caused by a variety of algae, including many species of diatoms, desmids, and filamentous green algae.

The middle site runs from Highway 51 to Williamsburg. The river in this area is characterized by long continuous stretches of similar depth with a low pool/riffle ratio. The stream gradient in the middle study area is low (0.27%). Two small springs

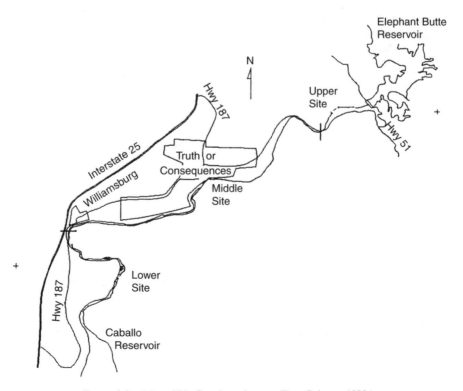

FIGURE 3.3.   Map of Rio Grande study area. (From Bristow, 1993.)

enter the river near Truth or Consequences, New Mexico. The first spring enters the river near the junction of Highways 187 and 51. The second enters downstream about 1 km (Bristow, 1993).

Where the river bends to the south, downstream from Williamsburg, the lower study area begins. In this portion of the river, the pool/riffle ratio (0.40:1.00) is greater than it is in the middle or upper study sites. In the upper and middle study sites the stream gradient is higher. Smaller substrate sizes and embeddedness are usually restricted to the upper portion of the lower study site (Bristow, 1993). These studies were made between the dates of October 1987, December 1987, and January 1988; September, October, and January in 1988–1989; and November, December, January, and February in 1989–1990 (Bristow, 1993).

The basis of the primary production estimation was a combination of values from natural and artificial substrates. For estimates of secondary production of benthic invertebrates the size frequency method was used. To estimate tertiary production of fish, biomass was used to estimate instantaneous growth. Biomass was estimated by the Zippin depletion method. Estimations of growth were made from scales and otoliths (Bristow, 1993).

Fish samples were taken using nets and seining in a blocked stretch of the lower site on October 3, 1987. Seining was done four times and each catch was kept separate to enable researchers to calculate a depletion estimate on the fish (Bristow, 1993).

Electrofishing using an 8-ft john-boat to float the electroshocking unit was employed in collecting all fish samples after October 3, 1987. A Smith-Root 500-watt portable generator was used to obtain ac power for the electric shocker. Researchers pulled the boat and waded in the river with the anode while the cathode trailed behind the boat (Bristow, 1993).

Example of Ecosystem: Elephant Butte Dam to Caballo Dam

These studies of riverine systems support the theory Eugene P. and Howard T. Odum that in each stable community there are four functional stages of nutrient and energy transfer, i.e., (1) detritus producers and primary producers, (2) detritivores and herbivores, (3) omnivores, and (4) carnivores.

It should be noted that the algae are quite diverse in species in this study area (Figure 3.3). The structure of the ecosystem in the Rio Grande can best be described between Elephant Butte Dam and Caballo Reservoir.

Primary Production. The primary production was carried out primarily by diatoms and various green algae, although some blue-green algae were also present. The structure of the ecosystem in the Rio Grande between Elephant Butte Dam and Caballo Reservoir is as follows. The primary producers were the green algae; *Spirogyra* and *Pediastrum*, and the diatoms, *Amphora*, *Anomoeoneis*, *Campylodiscus*, *Diatoma*, *Frustulia*, *Gyrosigma*, *Opephora*, *Surirella*, and *Synedra*.

The six-month periphyton production (dry weight) averaged 23.5 g m^{-2} with a mean biomass of 0.76 g m^{-2}. From Elephant Butte the organic loading of detritus and phytoplankton averaged 1075.56 g m^{-2} each six months.

Secondary Production. The primary consumers of the algae were the herbivores and omnivores. They were the crayfish, *Orconectes*, a carnivore; the mayflies,

Baetis and *Tricorythodes*, omnivores; and the caddisfly, *Hydropsyche*, an omnivore. There were also unidentified species of Trichoptera, Coleoptera, and Diptera. These species were either herbivores, herbivore-omnivores, or omnivores. Feeding on these algae were the omnivores, gizzard shad (*Dorosoma*) (order Clupeiformes), and the Chironomidae including *Ablabesmyia* (midge larvae).

The secondary consumers found in the fish were damselflies and dragonflies (Bristow, 1993). Mayflies, which are collector-gatherers, and Chironomidae, which are collector-filterers, dominated the invertebrate fauna. Mayflies belonging to the genera *Tricorythodes* and *Baetis* (omnivores) were present. About 85% of the total biomass of mayflies were species of the genus *Tricorythodes*. Also present in very low numbers but found in many of the fish stomachs was one species of the family Heptageniidae (Bristow, 1993).

The next most common were insects belonging to the order Diptera. They had an estimated mean dry weight productivity of 4.51 mg m^{-2} per day. *Ablabesmyia* sp., a carnivore, was the principal and most productive species. For all the Diptera the biomass estimate was 0.32 g m^{-2} per day (Bristow, 1993).

In the fish diets the most common invertebrate by weight was the midge, *Ablabesmyia* sp. The most common prey species found in the stomachs of piscivorous fish was the gizzard shad, *Dorosoma cepedianum*, an omnivore. An estimate of the stomach contents of all the fish was 93.5% plant and detritus material (primary producers), 6.4% herbivore-detritivores (secondary producers), and 0.1% carnivores (tertiary producers) (Bristow, 1993).

The photosynthetic efficiency was estimated to average 0.023%. Transfer efficiency averaged 1.12% from trophic level 1 (primary producers) to trophic level 2 (herbivore-detritivores). The transfer efficiency from herbivore-detritivores (trophic level 2) to omnivore-carnivores (trophic level 3) averaged 0.14%. Transfer efficiency was estimated to be 3.03% from trophic level 3 (omnivore-carnivores) to trophic level 4 (carnivore) (Bristow, 1993).

ECOSYSTEM SECONDARY PRODUCTIVITY:
ZOOPLANKTON, FISH, AND INVERTEBRATES

Zoobenthos Production
Zoobenthos biomass averaged 1.72 g m^{-2} and the six-month productivity of zoobenthos averaged 7.56 g m^{-2}. For the period of study, total fish biomass averaged 11.15 g m^{-2} and six-month production estimates averaged 0.59 g m^{-2} (Bristow, 1993).

Invertebrate Production
For chironomids the cohort production interval was assumed to be 35 days when the mean water temperature was 13°C. A temperature of 11.7°C is the minimum water temperature for emergence for *Ablabesmyia* sp. It was calculated from these data that *Ablabesmyia* sp. would produce nine generations per year in the study site, and three of these generations would be during the fall-winter study period (Bristow, 1993).

The regeneration time was estimated for various invertebrates. For Ephemeroptera it was assumed to be six months. It was assumed that Planaria would produce one generation per year and it is known that the trichopteran genus *Helicopsyche* produces one generation per year in the water temperature regime observed. Productions of Gastropoda, Coleoptera, Hemiptera, Nematoda, Odonata, Oligochaeta, and Diptera were considered negligible, as they contributed such a small part of the total biomass (Bristow, 1993).

Fish Production

The greatest biomass was in the common carp. Second was the gizzard shad. Of the fish found in November, the common carp, black bullhead, green sunfish, bluegill, longear sunfish, largemouth bass, and white bass were taken. The most common were the common carp, and the next most common was the white bass (Bristow, 1993).

The dry weight biomass of the common carp was 10,852.5 mg m^{-2}. For the gizzard shad it was 292.0 mg m^{-2}, and for the white bass it was 75.3 mg m^{-2}, which was much less abundant than the other two species. Among the fish species the gizzard shad had the highest six-month fall–winter production rate (355.7 mg m^{-2} dry weight), whereas the carp had a six-month production of 58.4 mg m^{-2}. In the fall–winter period the six-month production estimates were similar for white bass (7.9 mg m^{-2}) and largemouth bass, *Micropterus salmoides* (7.5 mg m^{-2}) (Bristow, 1993).

In the fall and winter the ratios of productivity to biomass varied from 0.01 for the carp to 3.13 for the largemouth bass. For fish species in which a large portion of the individuals were young fish, productivity/biomass ratios were highest (Bristow, 1993).

At the lower sites the total dry weight biomass estimates increased from 1756.6 mg m^{-2} in 1987–1988 (field season) to 15,000.2 mg m^{-2} in 1988–1989 (field season). It is interesting to note that in the field season of 1989–1990 the biomass decreased to 1095.2 mg m^{-2} (Bristow, 1993).

In the winter of 1987–1988 the white bass (carnivores) biomass estimates were highest (1371.2 mg m^{-2} dry weight). However, they decreased to 337.2 mg m^{-2} in 1988–1989 with a further decrease to 66.0 mg m^{-2} dry weight by the following field season of 1989–1990. Although the gizzard shad was only collected in the 1988–1989 winter period, they contributed the highest six-month production of 1185.6 mg m^{-2} dry weight and a second highest biomass of 949.8 mg m^{-2}. In the 1988–1989 field season, the gizzard shad collected were all age 0 (Bristow, 1993).

For all three sampling periods, the largemouth bass was present in low numbers. The mean sampling weight biomass estimates were 9.6 mg m^{-2} in both 1987–1988 and 1989–1990 seasons but were 30.16 mg m^{-2} in the 1988–1989 season. There was an increase in six-month dry weight production from 0.7 mg m^{-2} in 1988–1989 to 49.5 mg m^{-2} in the 1989–1990 field season (Bristow, 1993).

Oversummered rainbow trout, *Oncorhynchus mykiss*, were captured only in the 1987–1988 season (Bristow, 1993). Brown trout, *Salmo trutta* (carnivores), were caught the following year and their mean wet weight was about the same as the rainbow trout captured the year before (Bristow, 1993).

There was a dramatic decrease in fish diversity at the middle site from the winter of 1987–1988 to the winter of 1989–1990. After the high-water year of 1987, 10 species were captured at the middle site. In the 1989–1990 field season (a low-water year) only two species were collected. The mean dry weight biomass of all species was much lower in the 1987–1988 field season (8321.7 mg m^{-2}) than in the 1989–1990 field season (261,341.2 mg m^{-2}) (Bristow, 1993).

At the middle site, the number of fish collected was much greater than at the lower site during the high-water year of 1987–1988. During the low-water year of 1989–1990, more species were collected from the lower site. However, the middle site had the higher mean dry weight biomass for both years in which the two study sites were sampled (Bristow, 1993).

The fish species that were caught were rainbow trout, gizzard shad, threadfin shad, common carp, red shiner, bullhead minnow, black bullhead, yellow bullhead, white bass, white crappie, green sunfish, longear sunfish, bluegill, largemouth bass, and yellow perch. The food of the fish was mainly primary producers, 30.9% were herbivores, and first carnivores constituted less than 1% of all fish (Bristow, 1993).

The highest values for feeding on primary producers were carp (93.9%) and gizzard shad (95.2%). Feeding on herbivorous invertebrates were the red shiner, *Cyprinella lutrensis*; bullhead minnow, *Pimephales vigilax*; and the yellow perch, *Perca flavescens*. The fish with more than 99% herbivores in their stomachs were the green sunfish, *Lepomis cyanellus*; largemouth bass, and white bass (Bristow, 1993).

Two species of bullhead catfish and rainbow trout had the highest total percent of first carnivores in their stomachs. Measuring total consumption by weight from trophic level 3, yellow bullhead consumed 1.8%, black bullhead (*Ictalurus melas*) 0.3%, and rainbow trout 1.0% (Bristow, 1993).

GENERAL DESCRIPTION OF AQUATIC LIFE

Invertebrates Other Than Insects: Texas

Several oligochaete worms were collected from streams associated with the Rio Grande. They were *Pristina americana* (Cernosvitov), reported in 1937 from Presidio, Val Verde, and Zapata counties in streams associated with the Rio Grande (Harman et al., 1979). In streams from the Rio Grande in Brewster and El Paso counties and from Presidio, Webb, and San Patricio counties, *Nais variabilis* was collected (Harman et al., 1979).

Pristina synclites (Stephenson 1925) was found in the Rio Grande above Presidio in Presidio County. It was unusual to find it in the Rio Grande, as it bears no resemblance to the mountain streams in the region where it was found. The substrate where it was found was composed of thin silt deposits underlain by gravel. This species was uncommon at this locality (11 m^{-2}) and occurred sympatrically with the tubificid, *Limnodrilus udekemianus* (Harman et al., 1979).

Dero (Dero) digitata (Müller 1773) is recorded from the Rio Grande at Zaragosa International Bridge in El Paso County. It occurred in moderate numbers at the Zaragosa International Bridge site sympatrically with *Nais variabilis*, *Limnodrilus udekemianus*, and *L. hoffmeisteri*. The sandy substrate where *Dero digitata* was found is organically enriched by the discharges of sewage treatment plants. These sewage treatment plants contributed a major portion of the flow. In this area the water is typically very turbid, and the dissolved oxygen concentration often becomes depleted. *Dero digitata* was most common in organic silt deposits along the banks (Harman et al., 1979).

Limnodrilus hoffmeisteri was common at the Zaragosa International Bridge on the Rio Grande. *Limnodrilus udekemianus* Claparède 1862 was collected in the Pecos River in Pecos County, at two localities on the Rio Grande in El Paso County, and in the Rio Grande in Presidio County.

Insect Fauna: Texas

The Big Bend region embraces Presidio, Brewster, and Jeff Davis counties. The dragonfly, *Stylurus intricatus* (Hagen), was found in Hot Springs on the Rio Grande in Brewster County. One female was found on July 12, 1930, and one male on October 4, 1935 (Gloyd, 1958). One female *Hetaerina americana* (Fabricius) (family Calopterygidae) was found on October 4, 1935 in Hot Springs on the Rio Grande in Brewster County. Also found in Hot Springs on the Rio Grande in Brewster County were *Argia immunda* (Hagen): one male, July 12, 1930; nine males and two females on October 4, 1935; and *Argia moesta* (Hagen): four males and two females, July 12, 1930; and one female, October 4, 1935 (Gloyd, 1958).

Enallagma civile (Hagen) was taken from Hot Springs on the Rio Grande (Brewster County). Four males were collected October 4, 1935 (Gloyd, 1958). Also collected were specimens of *Telebasis salva* (Hagen): eight males and two females on October 4, 1935 (Gloyd, 1958).

Rio Grande Texas Sections: Study of Macroinvertebrates

Davis (1980) divided the Texas Rio Grande into three sections in his study of insects: the Bolson section, which extends from New Mexico to Fort Quitman; the Presidio section, from Fort Quitman to 50 km below Presidio; and the Canyon section. The upper Canyon section extends from 50 km below Presidio to La Linda, and the lower Canyon section is below La Linda (Davis, 1980).

Bolson Section. In the Bolson section at El Paso, the Oligochaetes were represented by one taxon, which was not very numerous. The Diptera were represented by seven taxons, none of which were very common. The mayflies were represented by two taxons, the Odonates by two taxons, and the Pelecypoda by a single taxon that was very common. The total number of individuals was 127 and the substratum sampled was 2 m² (Davis, 1980).

In the Bolson section at Zaragosa International Bridge (river km 24.5; 1118 m above mean sea level) the diversity was lowest and the redundancy and standing crop were the highest among the various areas studied. Only 16 taxa were collected, four of which were Oligochaetes, Psychodidae, and *Chironomus*. Together they made up 99% of the macrobenthos. In this reach there was severe environmental degradation caused by sewage discharge, which elevated the turbidity, ammonia nitrogen, and total phosphorus as well as total organic matter and oxygen depletion. The degradation also caused a buildup of fine sediments, which prevented the leaching of organic matter, oil and grease, and heavy metals (particularly lead). Degradation was indicated by the dominance of the tubificids, *Limnodrilus hoffmeisteri* and *L. udekemianus*. These two taxons comprised more than 80% of the macrobenthos on four sampling dates (Goodnight and Whitley, 1960) and had the anomalously low mean diversity value of 0.55 (Davis, 1980; Wilhm, 1967).

Presidio Section.   In the Presidio section, the upper Presidio station is 10.5 km above Presidio (river km 462.4; 785 m above mean sea level). The mean diversity was 1.29 and the total number of taxa were 18. This was comparable to the number of taxa found at El Paso. Five taxa (*Cheumatopsyche*, *Simulium* (blackflies), *Dicrotendipes* (chironomids), *Cricotopus*, and Oligochaeta) comprised 91% of the macrobenthos. Several environmental factors contributed to the suppression of diversity: (1) elevated dissolved solids caused by high evaporation/transpiration and tailwater inflow; (2) excessive turbidity and lack of habitat diversity caused by the predominance of fine sediments; (3) extreme physicochemical fluctuations caused by consistently low flow; and (4) possible chronic toxicity caused by low pesticide levels. DDT, DDD, and DDE were present in five species of fish (134 to 535 μg kg^{-1}, total DDT) and endrin was present in three species at a level of 14 to 37 μg kg^{-1}. Sediment concentrations were below detection limits (Davis, 1980).

Mean diversity of insects was 2.31 at the lower Presidio station (16.3 km below Presidio; river km 497.2; 771 m above mean sea level). This was more than at upper Presidio. A total of 34 taxa were collected, which was the highest among stations in the Chihuahuan biotic province. Five taxa (*Cheumatopsyche*, *Simulium*, *Thraulodes*, *Traverella*, and *Orthocladius*) comprised 71% of the macrobenthos (Davis, 1980).

Higher diversity than that which occurred upstream is probably due to the increased complexity of the substratum, resulting in a greater variety of microhabitats. The high discharge of the Rio Conchos inflow caused a dilution of the dissolved solids (particularly chlorides), turbidity decrease, and greater stability of physical and chemical characteristics. Greater predominance of gill-breathing organisms, particularly Ephemeroptera (mayflies), which occurred downstream from the Conchos confluence, indicated better water conditions. In the sediments the concentrations of DDT and DDE were 10.5 and 17.9 μg kg^{-1}, respectively. In two species of forage fish, the total DDT levels were 198 and 1284 μg kg^{-1}. The Rio Conchos was a primary source of these pollutants, as indicated by additional sediment analysis. Fifty kilometers below the confluence with the Rio Conchos, DDE was the only detectable sediment residue (2.5 μg kg^{-1}). Davis states that diversity may have been prevented from reaching its potential maximum as sensitive taxa are eliminated by the presence of pesticide residue in the sediments. The occurrence of deformed *Traverella* nymphs with shriveled legs and tumorous eyes was indicative of sublethal pesticide effects. In bottom sediments, persistent pesticide residues have been shown to damage benthic organisms despite low water concentrations (Davis, 1980; Wilson and Bond, 1969).

Canyon Section

Upper Canyon Section.   The Canyon section is the section below the lower Presidio section. The upper Canyon section includes Santa Elena Canyon (river km 600.5; 654 m above mean sea level). The mean diversity was 1.83, less than at lower Presidio, due primarily to the depaupered fauna consisting of 16 taxa. Five taxa (*Simulium* sp., *Traverella* sp., *Choroterpes mexicanus* sp., *Tricorythodes* sp., and *Thraulodes* sp.) comprised 82% of the macrobenthos. Two local factors appear detrimental to macrobenthic diversity: (1) high mean turbidity, with extremes during silt-laden rainfall runoff from alluvial tributaries and periodic high rates of sediment deposition, and (2) frequent modification of the substratum during spates from Terlingua Creek (Davis, 1980).

Lower Canyon Section. Below La Linda is the lower Canyon section. Here the river is incised to the hilly Cretaceous limestone of the Stockton and Edwards plateaus. With the increasing rainfall downstream, the cover of desert shrubs (mesquite, oak, and juniper) thickens. Along the stream margins are extensive stands of cane grass and Bermuda grass. The maximum river width is 100 m at Del Rio. The mean gradient is 0.7 m km^{-1}. The Rio Grande, Pecos, and Devil's rivers converge below Foster's Ranch to form Amistad Reservoir. This reservoir has a storage capacity of 26,250 ha. It was completed in 1968. In the lower Canyon section there were two stations, the Rio Grande at Foster's Ranch (river km 990.8; 333 m above mean sea level) and the Del Rio International Bridge station (river km 1119.2; 265 m above mean sea level) (Davis, 1980).

Foster's Ranch. At Foster's Ranch, which is bordered by rugged hills, the stream has loose and embedded cobblestones, and silt also forms part of the substrate. The mean insect diversity was 2.92 and the total number of taxa was 39. These numbers were higher than in the Chihuahuan biotic province. Six taxa (*Traverella, Smicridea, Tricorythodes, Cheumatopsyche, Thraulodes,* and *Hydroptila*) comprised 72% of the macrobenthos. The diversity was higher than upstream, and several factors contributed to this condition: increased springflow discharge that diluted the concentration of dissolved solids and buffered the aquatic environment against sudden chemical and physical changes. Fluctuations in turbidity were not as extreme as at Santa Elena Canyon, although the mean turbidity was high, caused by fluctuations in the velocity of the flow. Second, the complexity of the substratum increased the habitat diversity. A third factor was the physiographic and biotic diversity of the region and the existence of tributaries that permitted greater recruitment of peripheral species than in the more arid reaches, which had few tributaries, such as the Chihuahuan biotic province. The geography is important, as indicated by (1) higher diversity of taxa at Foster's Ranch compared with the diversity at lower Presidio and Santa Elena Canyon, despite the fact that the physical/chemical environments were relatively similar, and (2) the occurrence of several taxa common in adjacent tributaries which occurred only in this lower reach (Davis, personal observations). Examples of these taxa were the neotropical genera, *Limnocoris, Cryphocricos,* and *Leptohyphes.* These taxa reached their northernmost limits of distribution in this area of the Rio Grande. Common inhabitants of the Edwards Plateau streams to the east included members of the genera *Brechmorhoga mendax, Baetodes, Isonychia, Elsianus texanus, Microcylloepus, Stenelmis, Lutrochus luteus, Psephenus texanus, Hydropsyche,* and *Oecetis,* (Davis, 1980).

Del Rio International Bridge Station at Del Rio. Macrobenthic diversity was enhanced by marked improvement in water quality resulting from silt removal in Amistad Reservoir and freshwater inflow from Devil's River (<250 mg L^{-1} dissolved solids). The salinity of the Pecos River is offset by the reduction of dissolved solids (ca. 1600 mg L^{-1} dissolved solids). Also contributing to biodiversity was the high physicochemical stability promoted by Amistad Reservoir's buffering effects and persistently high flow; and the very complex substratum described above. For example, the Del Rio International Bridge station (river km 1119.2; 265 m above mean sea level; and below the confluence with the Pecos) has a mean diversity of 3.31, and the number of taxa was 47, which is also high. Several taxa were unique to Del Rio. Six taxa (*Protoptila, Cheumatopsyche, Hydroptila, Corbicula manilensis, Tricorythodes*, and *Cricotopus trifascia*) composed 69% of the macrobenthos. The mean standing crop of algae growth was large (1938 m^{-2}). At six other stations the apparent limited algal diversity was probably due to low light penetration and high turbidity. However, the nutrient concentrations did not appear to be limiting (Davis, 1980).

FISH FAUNA

Extinct Species
Five species of fish that once inhabited Texas waters are now extinct: *Notropis orca* (phantom shiner), *N. simus simus* (Rio Grande bluntnose shiner), *Cyprinella lutrensis blairi* (Maravillas red shiner), *Gambusia amistadensis* (Amistad gambusia) and *Gambusia georgei* (San Marcos gambusia). Three more have presumably been extirpated from the state: *Oncorhynchus clarki virginalis* (Rio Grande cutthroat trout), which occurred in McKittrick and Limpia creeks in west Texas, none of which have been caught there since; *Hybognathus amarus* (Rio Grande silvery minnow), which once inhabited the entire Rio Grande in New Mexico but now has a sharply limited range; and *Gambusia senilis* (blotched gambusia), which once lived in the Devil's River but now occurs only in the Rio Conchos basin in Chihuahua, Mexico. More than 20% of the remaining primary freshwater species appear to be in some need of protection (Hubbs et al., 1991).

Restricted Species
Astyanax mexicanus (de Filippi) (Mexican tetra) is probably restricted to the Pecos downstream from Dexter (south of Roswell). However, it did appear in the Rio Grande drainage but appears to be extirpated. Originally, this fish was restricted to the Rio Grande and Pecos drainages and the Nueces drainage in Texas, but the range has been extended, largely through its use as fish bait in the southwest and south-central states. Historical records indicate that these small fish that are intolerant of cold originally migrated upstream in the Rio Grande (Sublette et al., 1990).

    Cyprinidae. Hybognathus amarus (Girard) (Rio Grande silvery minnow) is extirpated in Texas and exists only in scattered Rio Grande locations in New Mexico. Freshwater. Extirpated (Hubbs et al., 1991).

    Salmonidae. Oncorhynchus clarki (Richardson) (cutthroat trout), was originally widespread in streams from Alaska to northern California eastward through the intermountain basins to the upper Missouri or Arkansas, Platte, Colorado, and Rio Grande systems. Current restricted range of this subspecies includes the headwaters of the Rio Grande and Pecos drainages, possibly the headwaters of the Canadian drainage. It is thought to have been present originally in at least Limpia and McKittrick creeks in Texas and possibly elsewhere in the Davis Mountains. Freshwater. Presumed extirpated (Hubbs et al., 1991).

Fish Fauna from Headwaters to Mouth: General Remarks
The fish fauna of the Rio Grande is varied and differs as one proceeds from headwaters to mouth. This change in fish fauna is correlated with water conditions and other characteristics.

At a site of the Rio Grande near Del Norte, Colorado, at an elevation of 7980 ft with a drainage basin of 1311 mi^2, the main fish collected was the fathead minnow (*Pimephales promelas*), which is an omnivore. At an elevation of 6050 ft, below Taos Junction Bridge near Taos, New Mexico, with a contributing drainage basin of 6527 mi^2, the fish taken were the carp, *Cyprinus carpio*; *Gila pandora* (Rio Grande chub); *Rhinichthys cataractae* (longnose dace); *Catostomus commersoni* (white sucker); *C. plebeius* (Rio Grande sucker); *Oncorynchus mykiss* (rainbow trout); *Salmo trutta* (brown trout); and *Micropterus dolomieui* (smallmouth bass). There were a total of eight species, three of which were omnivores. The total number of tolerant species was two.

At site 9M, the Rio Grande at Isleta, New Mexico, and at an elevation of 4896 ft with a drainage area of 14,537 mi^2, the fish collected during 1993 and 1994 were *Cyprinella lutrensis*, *Cyprinus carpio*, *Hybognathus amarus*, *Pimephales promelas*, *Platygobio gracilis* (flathead chub), *Rhinichthys cataractae*, *Carpiodes carpio*, *Catosomus commersoni* (white sucker), *C. plebeius* (Rio Grande sucker), *Ameiurus natalis* (yellow bullhead), and *Ictalurus punctatus* (channel catfish). Normally, *Hybognathus amarus* was found in northern New Mexico in the Rio Grande. Evidence suggests that it was one of the most abundant species, but it has not recently been reported from many portions of its former range (Carter, 1997b).

Fish of New Mexico
Anguilliformes
Anguillidae.   *Dorosoma cepedianum* was found in the Rio Grande south of the 35th parallel (roughly south of Albuquerque). The adult gizzard shad is primarily a bottom feeder feeding on detritus, ingesting large parts of floating material and phytoplankton. The juveniles also take Protozoa and copepods as well as Cladocera and various green and yellow-green algae (Sublette et al., 1990).

Cypriniformes.   *Agosia chrysogaster* Girard (order Cypriniformes), the longfin dace, was introduced into the Rio Grande downstream from Elephant Butte Reservoir, where it is now localized. Adults feed primarily on detritus but also take zooplankton, aquatic insects, and filamentous algae. It is an omnivore (Sublette et al., 1990).

Campostoma anomalum (Rafinesque), commonly known as the central stoneroller, has been introduced into the Rio Grande near Albuquerque and Truth or Consequences. It is rather widely distributed in the Rio Grande drainage system. This species prefers warm, shallow water with abundant aquatic vegetation. Spawning begins in spring or early summer when the water temperatures are between 13 and 27°C. It is primarily an herbivore, feeding diurnally on filamentous algae and diatoms but also taking detritus and aquatic insects from the periphyton assemblages that occur on rock surfaces. They are voracious feeders and their feeding can reduce the algae flora of a pool significantly (Sublette et al., 1990).

The red shiner, *Cyprinella lutrensis* (Baird and Girard), is a native of the Pecos and Rio Grande drainages. They avoid temperature extremes in winter and summer, consistently selecting water deeper than 20 cm with negligible flow and a pH of 7.1 to 7.4. Dissolved oxygen, turbidity, shade, and substrate are of relatively less importance on habitat selection of the red shiner (Sublette et al., 1990).

Catostomidae.   *Ictiobus bubalus* (Rafinesque), the smallmouth buffalo, is native to the Rio Grande in Sierra and Dona Ana counties (Sublette et al., 1990).

Moxostoma congestum (Baird and Girard), the gray redhorse, was formerly found in the Rio Grande downstream from Elephant Butte Reservoir. Currently, it occurs only in the Pecos River and Black River (a tributary of the Pecos). It has also been found in the Rio Grande in Texas (Sublette et al., 1990).

In New Mexico, *Carpiodes carpio* (Rafinesque), often called the river carpsucker, is found in the middle elevations. *Catostomus (Pantosteus) plebeius* Baird and Girard, commonly known as the Rio Grande sucker, is found in the Rio Grande and its tributaries primarily north of the 33rd parallel. It is believed that the population of this species is stable. It has been reported from other drainages. Populations of this small mountain sucker have declined throughout the middle Rio Grande region. They are listed in the Colorado Division of Wildlife (CDOW) as endangered on the upper Rio Grande ecosystem of Colorado (Langlos et al., 1994). This fish inhabits both riverine and lacustrine habitats. The Rio Grande sucker also occurs in Mexico (Sublette et al., 1990).

Cyprinidae.   *Cyprinus carpio* Linnaeus (common carp) is found in New Mexico in a wide variety of habitats. Sanchez (1970) reported this species spawning in the upper end of the Elephant Butte Reservoir. This species is often considered a pest because it uproots aquatic vegetation and feeds on the eggs of more desirable species (Sublette et al., 1990).

Gila pandora (Cope) (Rio Grande chub) is found in the Rio Grande drainage in Colorado. It is often associated with aquatic vegetation and is found in impoundments. It is a midwater carnivore feeding on zooplankton, aquatic insects, and juvenile fish. It sometimes takes a small amount of detritus. Thus it is an omnivore (Sublette et al., 1990).

Hybognathus amarus (Girard), often referred to as the Rio Grande silvery minnow, occurs in the Rio Grande downstream from Velarde and downstream from Abiquiu. It occurs only in the perennial sections of the stream and is sometimes found in irrigation canals between Cochiti and Socorro (Sublette et al., 1990).

Notropis jemezanus (Cope), known as the Rio Grande shiner, prefers large open rivers with laminar flow and a minimum of aquatic vegetation. It prefers gravel, sand, and rubble beds of rivers. These are sometimes overlain with silt (Sublette et al., 1990). *Notropis simus* (Cope) (bluntnose shiner) seems to be endemic to the Rio Grande and Pecos River drainages. Another species found in these areas is *N. stramineus* (Cope) (sand shiner) (Sublette et al., 1990).

Tolerant of turbid waters with low oxygen and high temperatures is *Pimephales promelas* Rafinesque, known as the fathead minnow. In New Mexico, spawning occurs in the spring and summer. This species feeds in soft bottom mud and takes a variety of items from algae and plant fragments to insect larvae and microscopic Crustacea. Thus it is an omnivore (Sublette et al., 1990).

Native to the middle and upper elevations of the Rio Grande is the longnose dace, *Rhinichthys cataractae* (Valenciennes). It has been introduced in the lower portions of the Rio Grande in the areas that have been modified by impoundments with tailwaters that simulate conditions found at higher elevations (Sublette et al., 1990).

Siluriformes

Ictaluridae. *Ictalurus (Ameiurus) natalis* (Lesueur), the yellow bullhead, now seems to be fairly well established in the Rio Grande and other streams. *I. (Ictalurus) furcatus* Lesueur, the blue catfish, is native to the Rio Grande downstream of Bernalillo County and the Pecos River downstream from Puerto de Luna. Populations of *I. (Ictalurus) furcatus* Lesueur, seem to be diminishing in the Rio Grande in Texas. *I. (Ictalurus) punctatus* (Rafinesque), the channel catfish, has been found in both the Pecos River and the Rio Grande (Sublette et al., 1990).

The flathead catfish, *Pylodictis olivaris* (Rafinesque), is native to the Rio Grande. Formerly, this species was believed to be distributed in the northern part of the Rio Grande. This is not so now (Sublette et al., 1990).

Lucania parva (Baird) (rainwater killifish) is an exotic species and has been found in the Rio Grande drainage as well as in various waters of the Rio Grande floodplain downstream from Elephant Butte Reservoir (Sublette et al., 1990).

Esocidae. Another fish in New Mexico is *Esox lucius* Linnaeus (northern pike). It inhabits rivers with slow current and abundant aquatic vegetation or other cover. It is a cool-water fish, and they often die off when the water temperatures reach 32°C (Goddard and Redmond, 1978). This fish spawns in late March or sometimes during later spring months. It typically spawns over vegetation, optimally short grasses or sedges, occasionally on mud bottoms, and in water temperatures of 6.2 to 18.5°C. The peak spawning is usually at 8.4°C. This species was introduced into the state in the 1960s. Populations are known from Navajo (San Juan River in Colorado), Elephant Butte, Cochiti, Sumner (Pecos River), and Conchas (Canadian River) reservoirs and the Rio Grande upstream of Cochiti Dam (Sublette et al., 1990).

Salmoniformes

Salmonidae. The cutthroat trout (*Oncorhynchus clarki virginalis*) is native to the Rio Grande. It is an opportunistic feeder, and terrestrial insects are often common in its diet. It also feeds on zooplankton and Crustacea. Luecke (1986) noted that when *Oncorhynchus clarki henshawi* is less than 6 cm long, it prefers *Daphnia pulex*, whereas when it is larger it feeds on benthic organisms. In the cooler headwaters, trout matures at a smaller size than at lower headwaters (Sublette et al., 1990).

O. clarki occurs in the Rio Grande drainage of New Mexico. *O. clarki virginalis*, the state fish of New Mexico, and Gila trout, *O. gilae*, represent the only extant native salmonids in New Mexico (Sublette et al., 1990).

Unfortunately, most of the streams occupied by Rio Grande cutthroat have been affected by overgrazing by livestock. Limited vegetation in the watershed has led to an alteration of the nutrients and sediment load. Lack of productive riffle areas also affects the trout's survival because it needs these areas for spawning. It also needs undercut banks to escape predators. These are sometimes rare in the Rio Grande in New Mexico (Sublette et al., 1990).

Gambusia affinis (Baird and Girard), the mosquitofish, is native to the lower elevations of the Rio Grande. It was introduced widely in the southwestern United States as well as in other temperate and tropical waters of the world. It is often used as a predator for controlling mosquitos (Sublette et al., 1990).

Poecilia (Mollienesia) latipinna (Lesueur), the sailfin molly, has been introduced along the lower stretches of the Rio Grande in New Mexico (Sublette et al., 1990).

Perciformes

Centrarchidae. *Lepomis (Chaenobryttus) cyanellus* Rafinesque, green sunfish, populations are established and stable in the Rio Grande drainage. *L. (Chaenobryttus) gulosus* (Cuvier), warmouth, is presumed to be exotic to the Pecos and Rio Grande drainages. Presently, it is confined to the Elephant Butte Reservoir in the Rio Grande. It is also present in the lower Pecos. *L. (Lepomis) macrochirus* Rafinesque, the bluegill, is common in the smaller, perennial, warm-water streams of New Mexico. However, it has been found in the Rio Grande, probably south of the 36th parallel. *L. (Lepomis) megalotis* (Rafinesque), the longear sunfish, was introduced into the Rio Grande and is presently found in Elephant Butte and Caballo reservoirs and in the Rio Grande (Sublette et al., 1990).

Micropterus dolomieui Lacepede (smallmouth bass) is most abundant in, or downstream of, major reservoirs in the middle elevation of the Rio Grande (Sublette et al., 1990).

Pomoxis annularis Rafinesque, white crappie, occur up to 200 mm in total length and have been observed in schools in Caballo Reservoir (Ozmina, 1965). This fish seems to feed mainly on zooplankton when it is small. Larger individuals also feed on insects and fish, although they remain planktivorous, feeding on zooplankton and other invertebrates (Maret and Peters, 1979; Sublette et al., 1990).

Pomoxis nigromaculatus (Lesueur), black crappie, is a midwater carnivore, feeding on insects, planktonic Crustacea and other invertebrates, and even small fish. It was introduced into the Elephant Butte Reservoir as well as the Caballo, Conchas (Canadian River), and Ute (Canadian River) reservoirs. This fish occurs in both the Rio Grande and the Pecos River.

Percithyidae. Of the Perciformes, *Morone chrysops* (Rafinesque), white bass, is known in Rio Grande downstream from Cochiti Reservoir. *M. saxatilis* (Walbaum), striped bass, occurs in the Elephant Butte Reservoir and has moved downstream into Caballo Reservoir along with the intervening section of the Rio Grande (Sublette et al., 1990).

Ambloplites rupestris (Rafinesque), the rock bass, has been found in the Rio Grande drainage in the vicinity of Elephant Butte and Caballo reservoirs. It is believed presently to be extirpated in the Rio Grande drainage, but it is in the Pecos drainage (Sublette et al., 1990).

Lepisosteidae. *Lepisosteus osseus* (Linnaeus) (longnose gar) is found from Quebec throughout the eastern United States southward to the Rio Grande drainage in Texas, New Mexico, and Mexico. Freshwater (Hubbs et al., 1991).

In Texas, *Lepisosteus spathula* Lacepede (alligator gar) is found in coastal streams from the Red River to the Rio Grande. Freshwater (Hubbs et al., 1991).

Fish of Texas

Although the following species are mainly found in Texas, some may also be found in New Mexico.

Anguillidae.   Texas records of *Anguilla rostrata* (Lesueur) (American eel) include specimens from the Red River to the Rio Grande and from most of the large river systems of the state. Estuarine (Hubbs et al., 1991).

Ophichthidae (Snake Eels).   *Myrophis punctatus* Lutken (speckled worm eel) is found in downstream sections of a number of river systems, including the Neches and the Rio Grande. Marine (Hubbs et al., 1991).

Cyprinidae.   *Ctenopharyngodon idella* (Valenciennes) (grass carp) is found widely scattered in Canadian, Red, Sabine, Trinity, and Rio Grande basins of Texas. Introduced from Asia. Freshwater (Hubbs et al., 1991).

Dionda episcopa Girard (roundnose minnow) occurs in shallow pools of low-gradient streams with an abundance of aquatic vegetation. It occurs in the Big Bend region of the Rio Grande and northern Mexico (Sublette et al., 1990).

Extrarius aestivalis (Girard) (speckled chub) occurs in the Rio Grande downstream of San Ildefonso (White Rock Canyon). Presently, this species seems to be extirpated in the Rio Grande (Sublette et al., 1990).

Gila pandora (Cope) (Rio Grande chub) inhabits limited areas of the Rio Grande and Pecos basins in New Mexico and Colorado. Freshwater. Threatened (Hubbs et al., 1991).

Macrhybopsis aestivalis (Girard) (speckled chub) occurs in Texas in streams from the Rio Grande to the Red River. Freshwater (Hubbs et al., 1991).

Notropis braytoni Jordan and Evermann (Tamaulipas shiner) is restricted to the Rio Grande and Rio Conchos basins in Texas and Mexico. It also occurs in the lower Pecos River. Freshwater (Hubbs et al., 1991).

Notropis buchanani Meek (ghost shiner) ranges from the lower Rio Grande and its Mexican tributaries northward to the Great Lakes. Inhabits large silt-laden streams. Freshwater (Hubbs et al., 1991).

Notropis chihuahua Woolman (Chihuahua shiner) is limited in Texas to small tributaries of the Rio Grande in the Big Bend region. Elsewhere, it occurs primarily in the Rio Conchos basin in Chihuahua, Mexico. Freshwater. Threatened (Hubbs et al., 1991).

Notropis jemezanus (Cope) (Rio Grande shiner) originally ranged throughout the Rio Grande basin. Declined in abundance in recent years and appears spottily distributed within the basin. Freshwater. Threatened (Hubbs et al., 1991).

Notropis orca Woolman (phantom shiner). Freshwater. Extinct (Hubbs et al., 1991).

Notropis simus (Cope) (bluntnose shiner). Freshwater. Extinct (Hubbs et al., 1991).

Notropis stramineus (Cope) (sand shiner) is found sporadically on the Edwards Plateau, in the Big Bend region of the Rio Grande, and along the Red River. Freshwater (Hubbs et al., 1991).

Pimephales vigilax (Baird and Girard) (bullhead minnow) was introduced into the upper Rio Grande basin (Hubbs et al., 1991). This species is known only in the Rio Grande downstream of the 33rd parallel (Sublette et al., 1990).

Platygobio gracilis (Richardson) (flathead chub), an extremely rare species in Texas, is known from the Canadian River in the Panhandle in Texas and also occurs in the Rio Grande in northern New Mexico. Freshwater (Hubbs et al., 1991).

Rhinichthys cataractae (Valenciennes) (longnose dace) inhabits a wide range of northern North America. Range extends south along the Rio Grande and the Pecos River in New Mexico, and it occurs into Texas throughout the Rio Grande about to Laredo. Freshwater (Hubbs et al., 1991).

Catostomidae. *Moxostoma congestum* (Baird and Girard) (gray redhorse). Rio Grande drainage. Freshwater (Hubbs et al., 1991).

Characidae. *Astyanax mexicanus* (Filippi) (Mexican tetra) is native to the Rio Grande and possibly the Nueces River drainages. Freshwater (Hubbs et al., 1991).

Ictaluridae. *Ictalurus lupus* (Girard) (headwater catfish) is native to the Pecos and Rio Grande basins of Texas and New Mexico. Freshwater. Special concern (Hubbs et al., 1991).

Ictalurus punctatus (Rafinesque) (channel catfish) is widespread east of the Rocky Mountains in temperate North America. Ranges throughout the upper Rio Grande and Pecos basins. Freshwater (Hubbs et al., 1991).

Ictalurus sp. (Chihuahua catfish) is restricted to the Rio Grande basin from New Mexico south through Texas and into Mexico as far as Rio San Fernando. Freshwater. Special concern (Hubbs et al., 1991).

There is a report of *Noturus gyrinus* (Mitchill) (tadpole madtom) from the Rio Grande in Webb County that may be the result of an introduction. Elsewhere, this species ranges widely east of the Rocky Mountains, except in upland streams draining the Appalachian mountain chain. Freshwater (Hubbs et al., 1991).

Cyprinodontidae. *Cyprinodon variegatus* Lacepede (sheepshead minnow) inhabits primarily coastal waters on the Atlantic and Gulf coasts from Maine south through the Gulf of Mexico and the Caribbean to Venezuela. Sometimes extends considerable distances upstream in coastal streams, especially in the Rio Grande. Estuarine (Hubbs et al., 1991).

Fundulus zebrinus Jordan and Gilbert (plains killifish) is abundant throughout the southern Great Plains of the United States. In Texas it occurs widely in the western part of the state. As an introduced population it occurs in the Rio Grande and some tributaries in and near Big Bend National Park. Freshwater (Hubbs et al., 1991).

Lucania parva (Baird and Girard) (rainwater killifish) is native to coastal waters from Massachusetts to Tampico, Mexico. Occurs in Pecos River, Leon Creek (San Antonio area), and in Falcon Reservoir in the Rio Grande basin. Estuarine (Hubbs et al., 1991).

Poeciliidae. *Gambusia speciosa* Girard (Mexican mosquitofish) occurs primarily in Mexico, occupying streams and tributaries to the Rio Grande and more southern drainages. Freshwater (Hubbs et al., 1991).

The native range of *Poecilia formosa* (Girard) (Amazon molly) in Texas is the lower Rio Grande. Freshwater (Hubbs et al., 1991).

Poecilia latipinna (Lesueur) (sailfin molly) is known from numerous inland localities, primarily in spring-fed central Texas headwaters and in the lower Rio Grande. Estuarine (Hubbs et al., 1991).

Syngnathidae. *Microphis brachyurus* (Bleeker) (opossum pipefish) is found in Texas only in the lowermost reaches of the Rio Grande in Cameron County, although this species is widespread throughout the brackish waters of Central America, the Antilles, and scattered localities along the eastern Gulf and Atlantic coasts of the United States. Estuarine. Special concern (Hubbs et al., 1991).

Centropomidae. *Centropomus parallelus* Poey (fat snook) is known in Texas only from the lower Rio Grande near the confluence with the Gulf of Mexico. Estuarine. Special concern (Hubbs et al., 1991).

Centrarchidae. A number of introductions of *Lepomis humilis* (Girard) (orangespotted sunfish) have occurred into various systems as far south as the Rio Grande basin. Freshwater (Hubbs et al., 1991).

Lepomis microlophus (Gunther) (redear sunfish) ranges throughout most of the southeastern United States. It is native to the eastern two-thirds of Texas from the Red River to the Rio Grande. This sunfish has been widely transplanted throughout the state. Freshwater (Hubbs et al., 1991).

Pomoxis annularis Rafinesque (white crappie) occurred naturally in the eastern two-thirds of Texas, but introduced populations now may be found statewide except in the upper Texas portions of the Rio Grande and Pecos basins. Freshwater (Hubbs et al., 1991).

Percidae. *Etheostoma gracile* (Girard) (slough darter) occurs in streams throughout the Gulf Coastal Plain and is found in Texas from the Rio Grande to the Red River. The Rio Grande records are from Jordan and Evermann (1896) and from one recent collection (Chaney and Pons, 1989). Freshwater (Hubbs et al., 1991).

Perca flavescens (Mitchill) (yellow perch) is an introduced species in Texas, and has established breeding populations only in the Rio Grande near El Paso, in Meredith Reservoir on the Canadian River, and in Greenbelt Reservoir on the Salt Fork of the Red River, despite being introduced into many other waters in the state. Freshwater. Introduced (Hubbs et al., 1991).

Cichlidae. *Cichlasoma cyanoguttatum* (Baird and Girard) (Rio Grande cichlid) is native to the United States and Texas only in the Rio Grande and Pecos drainages. This species is also native to northeastern Mexico. Freshwater (Hubbs et al., 1991).

Tilapia aurea (Steindachner) (blue tilapia) is native to the Middle East and along the Mediterranean coast of North Africa. This aquacultural species has been introduced into Texas and has become established in the Rio Grande, San Antonio, Guadalupe, and parts of the Colorado River drainages. Most successful establishments are in areas without extremely cold winter water temperatures (e.g., the lower Rio Grande basin and reservoirs heated by power plant effluents). Freshwater. Introduced (Hubbs et al., 1991).

Mugilidae. *Agonostomus monticola* (Bancroft) (mountain mullet) has been found considerable distances upstream in various Texas streams from the Trinity to the Rio Grande; it is a common inhabitant of the lower Rio Grande. Estuarine (Hubbs et al., 1991).

Eleotridae. *Gobiomorus dormitor* Lacepede (bigmouth sleeper) is found in southern Florida and Texas and south through the Gulf of Mexico and the Caribbean to South America. This species inhabits the coastal regions of southern Texas. It is quite common in the lower reaches of the Rio Grande. Estuarine (Hubbs et al., 1991).

Gobiidae. *Awaous tajasica* (Lichtenstein) (river goby) is known in Texas only from the Rio Grande in Hidalgo and Willacy counties. Estuarine. Special concern (Hubbs et al., 1991).

Bathygobius soporator (Valenciennes) (frillfin goby) is found inhabiting most bays and estuaries in southern Texas. Also found frequently in the lower reaches of coastal streams, especially in the lower Rio Grande. Marine (Hubbs et al., 1991).

Gobionellus atripinnis (Gilbert and Randall) (blackfin goby) is a coastal species known only from a few records in southern Texas and northern Mexico. Originally described from lower Rio Grande. Estuarine. Endangered (Hubbs et al., 1991).

Fish Fauna: Falcon Reservoir to Mouth
In this region there were three principal tributaries that entered the Rio Grande, two from Mexico and one from the United States: Rio Alamo, Rio San Juan, and Los Olmos Creek. The area of the Rio Grande that was studied by Robert J. Edwards and Salvador Contreras-Balderas is located between the Falcon Reservoir downstream to the mouth, locally known as Boca Chica (Edwards and Contreras-Balderas, 1991).

In this stretch of the river, three dams are located. The uppermost one is Falcon Dam, the next is Anzalduas, and the farthest downstream is Retamal. The floodwaters are regulated by the latter two dams. Edwards and Contreras-Balderas (1991) state that the water would be released into the floodway systems in the United States and Mexico whenever the flow exceeds $538.1 \text{ m}^3 \text{ s}^{-1}$ (as measured in the channel between Brownsville and Matamoros) (Edwards and Contreras-Balderas, 1991). Flow did not exist in this area during the severe drought of the 1950s and does not at present when the river no longer flows into the Gulf of Mexico.

In this section Edwards and Contreras-Balderas obtained 104 species of fish. Their collections constitute the first record from the Rio Grande for more than half of these species. The composition of this fauna was mostly freshwater-restricted types (such as cyprinids and centrarchids) as well as certain species more commonly associated with estuaries or brackish-water environments. They found that 20 species inhabit most sections (or at one time inhabited most portions) of the Rio Grande from the area that is now Falcon Reservoir to the mouth. These are *Dorosoma cepedianum, D. petenense, Astyanax mexicanus, Cyprinus carpio, Notropis jemezanus, N. orca, Hybognathus amarus, Hybopsis aestivalis, Carpiodes carpio, Ictalurus furcatus, Strongylura marina, Cyprinodon variegatus, Fundulus grandis, Poecilia latipinna, P. formosa, Gambusia affinis, Menidia beryllina, Membras martinica, Aplodinotus grunniens,* and *Mugil cephalus. Notropis orca* was recorded throughout the entire length of the Rio Grande in Texas and Mexico before the 1940s and therefore was included in this list (Edwards and Contreras-Balderas, 1991).

Most portions of the upstream freshwater environments (segments A to C) (Figure 3.1) were commonly inhabited by 14 species. These species include *Atractosteus spatula, Lepisosteus osseus, Pimephales vigilax, Notropis amabilis, N. braytoni, N. lutrensis, Ictalurus punctatus, Micropterus salmoides, Lepomis gulosus, L. cyanellus,*

L. macrochirus, Pomoxis annularis, Cichlasoma cyanoguttatum, and *Oreochromis (= Tilapia) aureus.* From one stream segment only were the following six species captured. *Lepisosteus oculatus, Notropis buchanani,* and *Moxostoma congestum* were taken only in segment B above Anzalduas Dam. From segment C of the stream were obtained *Carassius auratus, Selene vomer,* and *Awaous tajasica. Awaous tajasica* was very rare in the Rio Grande. It apparently reached the northern edge of its distribution in this stream (Edwards and Contreras-Balderas, 1991).

Notemigonus chrysoleucas, Lepomis megalotis, and *L. auritus* seem to be restricted to segment A of the Rio Grande, Falcon Reservoir. Found in segments A and B were *Morone chrysops* and *Lepomis microlophus.* Historically, *Lucania parva* is known from both segments A and C. On two separate occasions, hybrids between *Lepomis microlophus* and *L. macrochirus* were found in the Falcon Reservoir. From a numerical standpoint, *Dorosoma cepedianum, D. petenense, Notropis lutrensis, Pimephales vigilax, Strongylura marina, Fundulus grandis, Poecilia latipinna, P. formosa, Gambusia affinis, Menidia beryllina, Lepomis macrochirus, Cichlasoma cyanoguttatum,* and *Oreochromis aureus* dominate the upstream fauna (Edwards and Contreras-Balderas, 1991).

Fish Fauna: Brownsville to Mouth
The downstream fish assemblage was originally comprised of upstream fish, such as the cyprinids, *Hybopsis aestivalis, H. amarus,* and *Notropis jemezanus;* a tetra (*Astyanax mexicanus*); and a sucker (*Cyprinus carpio*). Also, several brackish-water forms, such as two species of *Dorosoma,* the poeciliids, the cyprinodontids, and a number of more marine forms, such as sciaenids and gobiids, compose part of the downstream fauna (Edwards and Contreras-Balderas, 1991). These are fish that range widely in the Rio Grande.

The downstream fauna of stream segment D contains many more common upstream estuarine types (such as *Dorosoma*), many other estuarine and coastal species (such as sciaenids, gobiids, syngnathids), and a surprising number of marine species. This lowermost section of the Rio Grande is being heavily utilized as a nursery ground or as a spawning ground, as indicated by the large number of young individuals. Occasional "waifs" were found in the collections. They were specimens of *Histrio histrio, Platybelone argulus,* and *Epinephelus nigritus.* It is not known how common these species are in the Rio Grande. The species that once were found in segment D of the lower river but are not found or are rarely found in this segment of the river today include *Astyanax mexicanus, Cyprinus carpio, Poecilia formosa, Gambusia affinis,* and *Dormitator maculatus* (Edwards and Contreras-Balderas, 1991).

In segments as far upstream as B and C, a few estuarine species were found: *Agonostomus monticola, Gobiomorus dormitor,* and *Gobiosoma bosci.* The presence of the first two species is well documented from the Rio Grande and surrounding coastal areas. In both segments C and D, *Gobiomorus dormitor* appears frequently. They are in catches of sports fishers. In segment C, two other residents that are primarily coastal were found. They were *Selene vomer,* which was captured in 1853, and *Gobiosoma bosci* taken from sampling stations in segment C and near the mouth from 1981 to 1989. The absence of *S. vomer* in current collections indicates that this species was not a year-round resident of this stream segment. The presence of this species from segment C probably represents a natural upstream extension of the coastal populations of this form (Edwards and Contreras-Balderas, 1991).

These data indicate that there were originally two fish faunas in the lower Rio Grande, referred to as "upstream" and "downstream" faunas. Today, these two faunas are also found. However, there has been a general decline of some of the species throughout the lower Rio Grande in the last 100 years. These include *Hybopsis aestivalis, H. amarus, Notropis jezemanus, N. orca*, and possibly *Ictalurus furcatus*. In one or more of the upstream segments (A to C) some species seemed to be less abundant (or absent). They were *Notropis buchanani, N. amabilis, Moxostoma congestum, Aplodinotus grunniens*, and *Mugil cephalus*. Possibly due to competition with rather large populations of needlefish, *Strongylura marina*, certain fish appear to be declining in abundance. They were the gars (*Atractosteus spatula, Lepisosteus oculatus*, and *L. osseus*). In oxbow lakes and irrigation supply reservoirs throughout the lower Rio Grande area, gars continue to be quite common. Large populations of *Agonostomus monticola*, particularly in the areas below Anzalduas Dam, may be causing the decline in *Mugil cephalus* because of competition (Edwards and Contreras-Balderas, 1991).

Some species seem to be increasing in some portions of the Rio Grande: *Pimephales vigilax, Strongylura marina, Fundulus grandis, Cyprinodon variegatus, Agonostomus monticola*, and *Oreochromis aureus*. Whether or not needlefish and mountain mullet (*Strongylura marina* and *Agonostomus monticola*) were originally members of the fish fauna in the Rio Grande (only rarely captured or documented) is not clear. In the Rio Grande, both these species are now common members of the fish community. Changes in the salinity regimes in the upstream sections of the stream may be associated with the increase in abundance of cyprinodontids. Except following major storms, salinities in stream segments B and C were consistently in the range of 0.5 ppt. It may be that this elevated salinity provided environmental conditions for the successful colonization of these species. In 1975, the blue tilapia (*Oreochromis aureus*) was first found in the Rio Grande, but at this time, the species was restricted to the region above Falcon Dam. It was in a series of collections in 1972 from Falcon Reservoir that the blue tilapia was first captured in the lower Rio Grande (Edwards and Contreras-Balderas, personal observations). In most recent collections taken upstream from the Brownsville, Texas and Matamoros, Tamaulipas area, *Oreochromis aureus* was found to be the dominant perciform (and often the dominant taxon). Despite a massive reduction in the number of species during a winter die-off in late 1983, this species is increasing in abundance in this region (Wood, 1986). It appears to be able to colonize habitats in a more generalized fashion than nearly any other species found in the Rio Grande system. In some areas, the only species of fish present were *Oreochromis aureus* and *Cyprinodon variegatus*, whereas in other sites the only fish taken was *O. aureus*. These three species seem to be generalists in their food habits. Their greatest impact appeared to be on the native cichlid, *Cichlasoma cyanoguttatum* (Wood, 1986). The change in flow regime within the Rio Grande basin may be the cause of the increase in abundance of *Pimephales vigilax*. Prior to the construction of the dam, the present-day stream channel may have had reduced silt deposits. Prior to dam construction, relatively large floods scoured the stream periodically. Local increases in siltation have probably been caused by the present dams, which have both moderated peak flows and restricted stream gradients (Edwards and Contreras-Balderas, 1991).

In segment B in the 1850s no collections were taken. However, several changes in the ecological composition of the fish fauna may have resulted and are still apparent, especially in segment D of the stream. The upstream fauna today is essentially freshwater in its affinities, as it was originally. However, the species have changed somewhat. Estuarine species seem to be increasing in stream segment C, which is interpreted as a response of the fish to rising salinities in the area. By 1991 the abundance of centrarchids and cyprinids appeared to be much lower than during prior recorded surveys. It appears that restricted freshwater species have been eliminated and replaced by estuarine and marine forms in the lowermost stream segment (D). The change to the species presently found in the area does not seem to be a sudden change from the original species composition in the lower segment. As freshwater species were eliminated, they appear to have been replaced by estuarine taxa, and later, these were replaced by taxa with more marine affinities. At the lowermost segment, five freshwater species were taken. However, it should be noted that three were found only following very heavy freshwater inflows associated with Hurricane Gilbert in October 1988. The finding of estuarine forms in stream segments B and C may indicate further faunal changes within these sections of the Rio Grande, similar to the changes that have already occurred in segment D (Edwards and Contreras-Balderas, 1991).

EXAMPLE OF FUNCTIONING ECOSYSTEM: CABALLO DAM TO DEL RIO

A study of the functioning ecosystem in the Rio Grande was made in the reach from Caballo Dam to the city of Del Rio (Table 3.3).

Vegetation Habitat. The primary producers were *Diatoma vulgare*, *Cocconeis pediculus*, *C. placentula*, *Fragilaria capucina*, and *Gomphonema olivaceum*. These were living in and among the vegetation. The detritivores were *Dugesia tigrina* and *Dero digitata* and the plecopteran *Perlomyia* sp. Detritivore-omnivores were *Nais variabilis* and *Pristina synclites*. Herbivores found were *Callibaetis* sp., a mayfly; the tricopterans *Nectopsyche* sp. and *Oecetis* sp.; Chironomids: *Polypedilum* nr. *illinoense*, *Polypedilum* spp., and *Sergentia* sp.; and the fish *Dionda episcopa*. Herbivore-omnivores were the molluscs *Physa virgata*, *Gyraulus circumstriatus*, *G. parvus*, and *Helisoma campanulata*. Other herbivore-omnivores were *Centroptillum* sp., *Helichus immsi* (a beetle), *H. suturalis*, *Helichus* spp., and the Orthocladiini *Cricotopus trifascia*, *Cricotopus (sylvestris gp.)* sp., *Cricotopus* spp., *Cricotopus* nr. *bicinctus*, *Glyptotendipes* sp. the Chironomid, and the fish *Pimephales promelas*.

The omnivores feeding on the vegetation or found in among vegetation were *Hydra* sp., *Ferrissia fragilis*; *Corbicula manilensis (fluminea?)*; the insect *Anax walsinghami*; the mayflies *Baetis* spp., *Baetodes* sp.; *Microcylloepus* sp. and *Lutrochus luteus*, elmid beetles. Other insects in among the vegetation were the omnivorous

TABLE 3.3.   *Species List: Rio Grande*

Taxon[a]	V	M	S	G	R	Site[c]	Source[d]
		Lentic		Lotic			
Division Chrysophyta							
Class Bacillariophyta							
Order Bacillariales							
Suborder Coscinodiscaceae							
Family Cyanodiscaceae							
* *Cyclotella meneghiniana*		X				MI	a
* *Melosira granulata*					X	MI	a
Stephanodiscus astraea		X				MI	a
v. *minutula*							
S. v. *hantzschii*		X				MI	a
Order Biddulphiales							
Family Biddulphiaceae							
* *Biddulphia laevis*		X				MI	a
Order Fragilariales							
Family Fragilariaceae							
* *Diatoma vulgare*	X	X				MI	a
* *Fragilaria capucina*	X					MI	a
Synedra ulna		X				MI	a
Order Achnanthales							
Suborder Achnanthinae							
Family Achnanthaceae							
* *Cocconeis pediculus*	X			X		MI	a
* *C. placentula*	X			X		MI	a
Order Naviculales							
Family Naviculaceae							
* *Caloneis amphisbaena*		X				MI	a
Navicula spp.						MI	a
Family Entomoneidaceae							
Entomoneis paludosa		X				MI	a
Entomoneis sp.		X				MI	a
Family Cymbellaceae							
Cymbella sp.							
Family Gomphonemaceae							
* *Gomphonema olivaceum*	X			X		MI	a
Order Epithemiales							
Family Epithemiaceae							
* *Epithemia sorex*		X				MI	a
* *Rhopalodia gibba*		X				MI	a
Order Bacillariales							
Family Nitzschiaceae							
* *Nitzschia acicularis*		X				MI	a
* *N. (Lanceolatae gr.)*		X				MI	a
* *Nitzschia* sp.		X				MI	a
Order Surirellales							
Family Surirellaceae							
* *Surirella brightwelli*		X				MI	a

TABLE 3.3.   (*Continued*)

Taxon[a]	V	Lentic		Lotic		Site[c]	Source[d]
		M	S	G	R		

Taxon[a]	V	M	S	G	R	Site[c]	Source[d]
Family Surirellaceae (*cont.*)							
* *Surirella* sp.		X				MI	a
Phylum Porifera							
Class Demospongiae							
Order Haplosclerida							
Family Spongillidae							
*o *Dosilia palmeri*		X			X	U	a
Phylum Cnidaria							
Class Hydrozoa							
Order Hydroida							
Family Clavidae							
Cordylophora lacustris							
Family Hydridae							
*o *Hydra* sp.	X				X	U,ZB	a,b
Phylum Platyhelminthes							
Class Turbellaria							
Order Tricladida							
Family Planariidae							
*d *Dugesia tigrina*	X	X				DR,U	a,b
Phylum Nemertea							
Prostoma rubrum						U	a
Phylum Mollusca							
Class Gastropoda							
Order Basommatophora							
Family Ancylidae							
*o *Ferrissia fragilis*	X	X		X		U	a
*o *Ferrissia* sp.		X		X		DR	b
Family Lymnaeidae							
*o *Lymnaea bulimoides*		X				U	a
*o *L. obtrussa*		X				U	a
*o *L. parva*		X				U	a
Family Physidae							
*ho *Physa virgata*	X	X			X	DR,FR,U	a,b
Family Planorbidae							
*ho *Gyraulus circumstriatus*	X	X				U	a
*ho *G. parvus*	X					U	a
*ho *Helisoma campanulata*	X	X				U	a
Class Bivalvia							
Order Unionoida							
Family Unionidae							
*o *Anodonta* sp.		X		X		U	a
Order Veneroida							
Family Corbiculidae							
*o *Corbicula manilensis* (*fluminea?*)	X			X		DR,EP,FR,U	a,b

(*continued*)

TABLE 3.3.   (*Continued*)

		Substrate[b]					
		Lentic		Lotic			
Taxon[a]	V	M	S	G	R	Site[c]	Source[d]
Family Sphaeriidae							
*o *Eupera cubensis*		X				FR	b
*o *Pisidium casertanum*		X				U	a
Phylum Annelida							
Class Oligochaeta							
Order Lumbriculida							
Family Lumbriculidae							
*d *Lumbriculus variegatus*		X				U	a
Order Haplotaxida							
Family Naididae							
*d *Dero digitata*	X					ZB	b
*do *Nais variabilis*	X	X				ZB,SE	b
Pristina americana							
*do *Pristina synclites*	X	X				UP	b
Family Tubificidae							
*do *Brachiura sowerbyi*		X				FR	b
*do *Limnodrilus hoffmeisteri*		X				ZB	b
*do *L. udekemianus*		X				EP,ZB,UP	b
Phylum Arthropoda							
Class Crustacea							
Subclass Malacostraca							
Order Amphipoda							
Family Gammaridae							
* *Gammarus pecos*		X				U	a
Order Decapoda							
Family Cambaridae							
*c *Orconectes causeyi*		X		X	X	U	a
*c *Procambarus clarkii*		X		X	X	U	a
Family Palaemonidae							
Class Insecta							
Order Odonata							
Suborder Zygoptera							
Family Calopterygidae							
*c *Hetaerina* sp.	X	X		X		DR,FR,LP	b
Family Coenagrionidae							
*c *Argia* sp.	X	X		X		DR,FR,LP,SE, S2,S5	a,b
*c *Enallagma* spp.	X	X				S1,S2,S4,S6	a
Hyponeura lugene						S5,S6	a
Suborder Anisoptera							
Family Aeshnidae							
*o *Anax walsinghami*	X					S1,S6	a
Family Gomphidae							
*c *Erpetogomphus* sp.				X	X	DR,FR,LP	b
*c *Gomphus externus*		X		X	X	DR,EP	b
*c *G. modestus*		X		X	X	ZB	b

TABLE 3.3. (*Continued*)

Taxon[a]	V	M	S	G	R	Site[c]	Source[d]
		Lentic		Lotic			
	V	M	S	G	R		
Family Gomphidae (*cont.*)							
*c *Gomphus* spp.		X		X	X	DR,EP,LP,ZB	b
Family Libellulidae							
*c *Brechmorhaga mendax*		X				DR,FR	b
*c *Tramea* prob. *onusta*	X	X				S1,S2	a
Order Ephemeroptera							
Suborder Schistonota							
Family Baetidae							
*o *Baetis* spp.	X	X		X		FR,SE,UP,S1, S5,S6,S7	a,b
*o *Baetodes* sp.	X			X		DR,FR	b
*h *Callibaetis* sp.	X					S1,S2,S5,S6	a
*ho *Centroptillum* sp.	X			X		DR,FR,LP,SE, ZB,S1,S2,S5, S6	a,b
o *Dactylobaetis mexicanus*		X				FR	b
o *Homoeoneuria* sp.				X		EP,ZB	b
*d *Pseudocloeon* sp.		X	X	X		DR,FR,LP,SE	b
Family Heptageniidae							
o *Heptagenia* prob. *elegantula*		X		X		S5	a
Family Leptophlebiidae							
o *Choroterpes mexicanus*		X		X		FR,LP,SE,UP	b
do *Thraulodes* sp.				X		DR,FR,LP,SE UP	b
o *Traverella presidiana*				X		FR	b
Traverella sp.				X		DR,FR,LP, SE,UP	b
Family Oligoneuriidae							
o *Isonychia* sp.				X		DR,FR	b
Suborder Pannota							
Family Caenidae							
o *Caenis* prob. *simulans*		X		X		S1,S2	a
o *Caenis* sp.		X				FR,LP,UP	b
Family Tricorythidae							
o *Leptohyphes* sp.				X		DR	b
o *Tricorythodes* sp.		X		X		DR,EP,FR,LP, SE,S6,S7	a,b
Order Plecoptera							
Family Nemouridae							
*d *Perlomyia* sp.	X	X		X		S7	a
Order Hemiptera							
Family Corixidae							
c *Graptocorixa* prob. *abdominalis*						S1,S2	a

(*continued*)

TABLE 3.3.   (*Continued*)

Taxon[a]	V	M	S	G	R	Site[c]	Source[d]
		Lentic		Lotic			
Family Naucoridae							
*c *Ambrysus* sp.	X	X			X	DR,FR,LP	b
*c *Cryphocricos* sp.					X	DR	b
*c *Limnocoris* sp.					X	FR	b
Family Notonectidae							
*c *Notonecta undulata*		X			X	S6	a
Family Saldidae							
*c *Saldula* sp.		X				S6	a
Family Veliidae							
*c *Rhagovelia choreutes*					X	S2,S4	a
Order Megaloptera							
Family Corydalidae							
o *Corydalus cornutus*		X		X		DR,FR,LP	b
Order Coleoptera							
Family Dytiscidae							
*oc *Dytiscus* sp.	X			X		S6,S7	a
*c *Laccophilus* sp.	X			X		S1,S2,S6	a
Suborder Polyphaga							
Family Dryopidae							
*ho *Helichus immsi*	X			X		LP	b
*ho *H. suturalis*	X			X		FR	b
*ho *Helichus* spp.	X			X		DR,FR,LP	b
Family Elmidae							
Elsianus texanus				X		DR	b
o *Heterelmis glabra*				X		FR,LP	b
*o *Microcylloepus* sp.	X					DR	b
o *Stenelmis* sp.				X		DR	b
Family Limnichidae							
*o *Lutrochus luteus*	X					DR,FR	b
Family Psephenidae							
Psephenus texanus						DR,FR	b
Order Trichoptera							
Family Glossosomatidae							
o *Protoptila* sp.						DR,FR,LP	b
Family Hydropsychidae							
o *Cheumatopsyche* sp.				X		DR,FR,LP,UP, ZB,S5,S6	a,b
o *Hydropsyche* sp.				X		DR,S5,S6,S7	a,b
o *Smicridea* sp.				X		DR,FR,LP,SE, UP	b
Family Hydroptilidae							
*o *Hydroptila* sp.	X					DR,FR,LP,S6	a,b
*o *Mayatrichia* sp.	X	X		X		FR,LP,SE	b
Family Leptoceridae							
*h *Nectopsyche* sp.	X					SE	b
*h *Oecetis* sp.	X			X		DR,FR	b
Order Lepidoptera							

TABLE 3.3.   (*Continued*)

Taxon[a]	Substrate[b]					Site[c]	Source[d]
	Lentic			Lotic			
	V	M	S	G	R		
Family Pyralidae							
Cataclysta sp.						DR,LP	b
Order Diptera							
Suborder Nematocera							
Family Ceratopogoninae							
*o *Palpomyia tibialis*		X		X		EP,ZB,SE,UP, ZB,S1,S2, S4, S5,S6,S7	b
Family Chironomidae							
Subfamily Pentaneurini							
c *Ablabesmyia* sp.		X		X		UP	b
*c *Conchapelopia* sp.		X		X		DR,UP	b
*c *Larsia* sp.				X		S2	a
*c *Zavrelimyia* sp.		X		X		S1	a
Subfamily Tanypodini							
ho *Tanypus* sp.						ZB	b
Subfamily Diamesinae							
o *Diamesa* sp.				X		S1,S2,S5	a
Subfamily Orthocladiinae							
Genus nr. *Rheorthocladius* spp.						S5	a
*o *Corynoneura* sp.	X	X		X		EP,UP,ZB	b
*o *Thienemanniella* sp.		X		X		DR,UP	b
Subfamily Orthocladiini							
Cardiocladius sp.						LP	b
*o *Chaetocladius* sp.	X			X		S2,S5	a
*ho *Cricotopus* nr. *bicinctus*	X					S1,S2,S4,S5	a
*ho *C. trifascia*	X	X	X			DR	b
*ho *Cricotopus* spp.	X	X	X	X		DR,LP,SE,S1	a,b
*ho *Cricotopus* (*sylvestris gp.*) sp.	X					S7	a
*o *Eukiefferiella* sp.		X		X		DR	b
*o *Hydrobaenus* sp.		X		X		EP,ZB	b
*o *Orthocladius* sp.		X				DR,FR,LP	b
unident. Orthocladiinae							
Orthocladiinae spp.						FR,SE	b
Subfamily Chironominae							
Tribe Chironomini							
*o *Chironomus* spp.	X	X		X		EP,ZB,S1,S2,S5	a,b
*c *Cryptochironomus fulvus*		X				LP	b
*o *Dicrotendipes* spp.		X				DR,LP,UP,S2	a,b
o *Endochironomus* sp.		X				S1,S2	a
*ho *Glyptotendipes* sp.	X	X		X		S1	a
o *Goeldochironomus* sp.		X				S6	a
*o *Paracladopelma* sp.		X				EP	b
*o *Paralauterborniella* sp.	X	X				LP	b
*h *Polypedilum* nr. *illinoense*	X	X				S1	a

(*continued*)

TABLE 3.3. (*Continued*)

Taxon[a]	Lentic			Lotic		Site[c]	Source[d]
	V	M	S	G	R		
Family Chironomidae (*cont.*)							
*h *Polypedilum* spp.	X					DR,EP,FR,LP SE,ZB	b
*o *Saetheria* sp.	X	X		X		EP,SE,UP	b
*h *Sergentia* sp.	X	X				S5	a
*o *Stenochironomus* sp.	X	X		X		LP,UP	b
Subfamily Pseudochironomini							
*o *Pseudochironomus* sp.	X	X		X		LP	b
Tribe Tanytarsini							
*o *Cladotanytarsus* sp.	X					FR,LP	b
Paratanytarsus (*Stylotanytarsus*)		X		X		S5	a
*o *Rheotanytarsus* sp.		X		X		DR,LP	b
*o *Tanytarsus* spp.	X	X		X		DR,LP,UP	b
Family Culicidae							
*o *Culex* sp. prob. *tarsalis*		X				S6	a
Family Simuliidae							
*o *Simulium* spp.	X	X		X		DR,FR,LP	a,b
Suborder Brachycera							
Family Stratiomyidae							
o *Euparyphus* sp.				X		DR	b
Family Tabanidae							
*c *Tabanus* sp.	X		X			DR	b
Phylum Chordata							
Subphylum Vertebrata							
Class Osteichthyes							
Order Anguilliformes							
Family Anguillidae							
*c *Anguilla rostrata*	X	X			X	C(h)	a
Order Clupeiformes							
Family Clupeidae							
*o *Dorosoma cepedianum*		X		X		BEB,C(h),C(l)	a,c
Order Cypriniformes							
Family Catostomidae							
*do *Carpiodes carpio*		X		X		BEB,C(h),C(l), CN,DA	a,c
*o *Catostomus commersoni*	X	X		X		BEB,C(h)	a,c
*o *C. plebeius*		X		X		BEB,C(h)	a,c
*o *Ictiobus bubalus*	X					BEB,C(h),C(l), DA	a,c
Order Characiformes							
Family Characidae							
co *Astyanax mexicanus*		X				C(h),CN,DA	a

TABLE 3.3. (*Continued*)

Taxon[a]	Substrate[b]					Site[c]	Source[d]
	Lentic			Lotic			
	V	M	S	G	R		
Family Cyprinidae							
*o *Carassius auratus*	X					BEB,C(h),CN, DA	a,c
*o *Cyprinus carpio*	X	X				BEB,C(h),C(l), CN,DA	a,c
*h *Dionda episcopa*	X	X				C(h)	a
*c *Gila nigrescens*	X	X				BEB,C(h)	a,c
*c *Hybognathus placita*					X	BEB,C(h)	a,c
o *Hybopsis aestivalis*				X		BEB,C(h),CN, DA	a,c
o *H. gracilis*				X		BEB	c
*o *Notemigonus crysoleucas*	X					C(h)	a
*o *Notropis jemezanus*		X				BEB,C(h)	a,c
*o *N. lutrensis*		X		X		BEB,C(h),C(l), CN,DA	a,c
o *N. proserpinus*						BEB	c
co *N. simus*		X				BEB	c
*ho *Pimephales promelas*	X	X				BEB,C(h)	a,c
Pimephales vigilax							
*c *Rhinichthys cataractae*					X	BEB,C(h),CN	a,c
*o *Tinca tinca*	X	X				BEB	c
Order Siluriformes							
Family Ictaluridae							
*c *Ictalurus furcatus*		X				BEB,C(h)	a,c
*c *I. melas*		X				BEB,C(h)	a,c
*c *I. natalis*		X				BEB,C(h),C(l)	a,c
o *I. punctatus*		X		X		BEB,C(h),C(l), CN,DA	a,c
*c *Pylodictis olivaris*		X		X		BEB,C(h),C(l), CN,DA	a,c
Order Salmoniformes							
Family Esocidae							
*c *Esox lucius*				X		BEB,C(h)	a,c
Family Salmonidae							
*c *Salmo clarki*				X		BEB	c
*c *S. gairdneri*				X		BEB,C(h)	a,c
*c *S. trutta*				X		BEB,C(h)	a,c
Order Cyprinodontiformes							
Family Poeciliidae							
*co *Gambusia affinis*	X	X				BEB,C(h),C(l), CN,DA	a,c
*co *G. gaigei*	X	X				CN	a
Order Perciformes							
Family Centrarchidae							

(*continued*)

TABLE 3.3.   (*Continued*)

Taxon[a]	V	M	S	G	R	Site[c]	Source[d]
		Lentic		Lotic			
Family Centrarchidae (*cont.*)							
*c *Micropterus dolomieui*				X		BEB	c
*c *Chaenobryttus gulosus*	X	X		X		BEB	c
*c *Lepomis cyanellus*	X	X		X		BEB,C(h),C(l)	a,c
*c *L. gulosus*	X	X				C(h),C(l)	a
*o *L. macrochirus*	X	X		X		BEB,C(h),C(l)	a,c
*c *L. megalotis*	X	X		X		BEB,C(h),C(l), CN,DA	a,c
*c *M. salmoides*	X	X		X		BEB,C(h),C(l), CN,DA	a,c
*c *Pomoxis annularis*	X					BEB,C(h),C(l)	a,c
*c *P. nigromaculatus*	X	X				BEB,C(h)	a,c
Family Percidae							
*c *Perca flavescens*		X		X		BEB,C(h),C(l), DA	a,c
*c *Stizostedion vitreum*				X		BEB,C(h)	a,c
Family Percithyidae							
*c *Morone chrysops*		X			X	BEB,C(h)	a,c
Family Sciaenidae							
*c *Aplodinotus grunniens*		X				C(h)	a

*Species found in the ecosystem described in the section "Example of Functioning Ecosystems."

[a] c, Carnivore; co, carnivore-omnivore; d, detritivore; do, detrivore-omnivore (devours organisms and all types of detritus; h, herbivore; ho, herbivore-omnivore; o, omnivore; oc, omnivore-carnivore.

[b] V, vegetation; M, mud; S, sand; G, gravel; R, rock.

[c] BEB, between Bernalillo and Elephant Butte Reservoir (middle Rio Grande valley, NM); C(h), below Caballo Dam, high flow (upper Rio Grande valley, NM); C(l), below Caballo Dam, low flow (upper Rio Grande valley, NM); CN, Canutillo (just south of Anthony, upper Rio Grande valley, TX); DA, Doña Ana, NM; DR, Del Rio (upper Rio Grande valley, TX); EP, El Paso (upper Rio Grande valley, TX); FR, Foster's Ranch (upper Rio Grande valley, TX); LP, lower Presidio (upper Rio Grande valley, TX); MI, municipal water plant intake, El Paso (upper Rio Grande valley, TX); S1, east drain, 2 mi south of Anthony, El Paso County, TX; S2, La Mesa Lake, 1.5 mi west of La Mesa, Doña Ana County, NM; S3, transient pond in an alkali swale, 3 mi southeast of McNary, Hudspeth County, TX; S4, Del Rio drain, $\frac{1}{2}$ mi northwest of Mesquite, Doña Ana County, NM; S5, Rio Grande at Buckle Bar Canyon, 4 mi west-northwest of Radium Springs, Doña Ana County, NM; S6, Rio Grande at Percha Diversion Dam spillway, 2 mi south of Caballo Dam, Sierra County, NM; S7, Las Animas Creek at I-25; $\frac{1}{2}$ mi south of the town of Caballo, Sierra County, NM; SE, Santa Elena Canyon (upper Rio Grande valley, TX); U, upper Rio Grande valley (Rio Grande between Caballo Dam, Sierra County, NM and American Diversion Dam, El Paso, TX); UP, upper Presidio (upper Rio Grande valley, TX); ZB, Zaragosa International Bridge (upper Rio Grande valley, TX).

[d] a, ANSP (1975); b, Davis (1980); c, Hansmann and Scott (1977).

trichopterans *Hydroptila* sp., *Mayatrichia* sp.; the omnivorous Chironomids: the Orthocladiinae *Corynoneura* sp., *Chaetocladius* sp., *Chironomus* spp., *Paralauterborniella* sp., *Saetheria* sp., *Stenochironomus* sp., *Pseudochironomus* sp., *Cladotanytarsus* sp., and *Tanytarsus* spp. Blackfly larvae were also present belonging to the genus *Simulium* spp. Omnivorous fish found in and among the vegetation were *Catostomus commersoni*; *Ictiobus bubalus*; the cyprinids *Carassius auratus, Cyprinus carpio, Notemigonus crysoleucas,* and *Tinca tinca*; and *Lepomis macrochirus.*

The carnivores found in among the vegetation were the Odonates *Hetaerina* sp., *Argia* sp., and unidentified species of *Enallagma*. Other carnivores found among the vegetation were *Tramea* prob. *onusta,* the stonefly *Ambrysus* sp., the coleopteran *Laccophilus* sp., and the horsefly *Tabanus* sp. Carnivore-omnivores found here were the coleopteran *Dytiscus* sp. and the fish *Gambusia affinis* and *G. gaigei*. The fish that were carnivores in among the vegetation were *Anguilla rostrata*; *Gila nigrescens*; and the centrarchids, *Micropterus salmoides, Chaenobrytus gulosus, Lepomis cyanellus, L. gulosus, L. megalotis, Pomoxis annularis,* and *P. nigromaculatus.*

Lentic Habitat. In areas where the current was relatively slow, such as along the edges of the stream or where it was almost absent, such as in pools and backwaters, the ecosystem was composed of many species. Primary producers living attached to the substrates were *Caloneis amphisbaena, Diatoma vulgare,* and *Rhopalodia gibba*. Floating near the surface of the substrate were species such as *Cyclotella meneghiniana*. Lying on the surface of the mud was *Biddulphia laevis*. Attached to the plants or debris was *Epithemia sorex*. Moving into or on the surface of the soft mud substrate were *Nitzschia acicularis, Nitzschia (Lanceolatae group),* and an unidentified species of *Nitzschia*. On the surface of the mud, particularly in pools but sometimes along the edges of the stream, were *Surirella brightwelli* and an unidentified species of *Surirella*.

In this habitat, where mud and sand are deposited and various bits of debris were present, was the detritivore, *Pseudocloeon* sp. Crawling over the detritus was the detritivore, *Dugesia tigrina*. Crawling over the debris, particularly in pools where mud was deposited, was the detritivore, *Lumbriculus variegatus*. Several detritivore-omnivores were found associated with the mud deposited along the edge of streams or in the pools. In this habitat were *Nais variabilis, Pristina synclites, Branchiura sowerbyi, Limnodrillus hoffmeisteri,* and *L. udekemianus*. Herbivore-omnivores in these habitats where the current was slow were *Physa virgata* and the planorbids, *Gyraulus circumstriatus* and *Helisoma campanulata.*

Living usually attached to debris were the sphaerids, *Eupera cubensis* and *Pisidium casertanum*, both omnivores. Also present in these lentic habitats were the omnivorous *Palpomyia tibialis, Corynoneura* sp., *Thienemanniella* sp., *Eukiefferiella* sp., *Hydrobaenus* sp., *Orthocladius* sp., *Chironomus* spp., *Dicrotendipes* spp.,

Paracladopelma sp., and *Paralauterborniella* sp. living in and among the debris in pools and where the current had been slowed. In these lentic habitat crawling in and among the soft silty mud were the omnivores *Saetheria* sp., *Stenochironomus* sp., *Pseudochironomus* sp., *Rheotanytarsus* sp., and *Tanytarsus* spp. The sponge, *Dosilia palmeri*, which is an omnivore, was found on the surface of the substrate (pebbles, sticks, etc.). Often, however, it was found on the surface of leaves of plants that had been deposited in the pools or occasionally growing on other algae. Omnivores found in this habitat that were attached to the substrate, such as leaves and rock surfaces, were *Ferrissia fragilis* and an unidentified species of *Ferrissia*. Also present were the omnivorous *Lymnaea bulimoides*, *L. obtrussa*, *L. parva*, and *Anodonta* sp.

The carnivorous *Conchapelopia* and *Zavrelimyia* sp. were also found in this habitat, feeding on various organisms. Crawling in and among the muddy sand were the herbivore-omnivores, *Cricotopus trifascia* and *Cricotopus* spp. In the soft muddy habitats were the omnivorous chironomids, *Chironomus* spp. Also in this lentic habitat were the herbivores, *Polypedilum* nr. *illinoense* and *Sergentia* sp. In these lentic habitats were also *Culex* prob. *tarsalis*, which is an omnivore, and unidentified species of the blackfly *Simulium*, an omnivore. These species are typically found where the current is moderately fast.

In these lentic habitats, particularly in pools or along the margins of the stream, were the herbivore, *Dionda episcopa*, probably feeding on bits of vegetation in these areas where the current was not too fast, and the herbivore-omnivore, *Pimephales promelas*. Also present was the detritivore-omnivore, *Carpiodes carpio*. This species is often found in fairly large pools. Several omnivorous fish were found in the pools and along the margins of the river. They were *Dorosoma cepedianum* and the catfish *Catostomus commersoni* and *C. plebeius*. The cyprinid, *Cyprinus carpio*, was feeding on the bottom of the pool. It is omnivorous. Other omnivores found in pool areas were *Notropis jemezanus* and *N. lutrensis*. The omnivorous *Tinca tinca* was also present, as was the omnivorous *Lepomis macrochirus*.

In these lentic habitats, often found hiding under rocks or debris or vegetation were the amphipod *Gammarus pecos* and the carnivorous crayfish *Orconectes causeyi* and *Procambarus clarkii*. Also present were the dragonfly *Hetaerina* sp., a carnivore, as is *Argia* sp. Other carnivorous Odonates in this habitat were *Enallagma* spp. Other carnivores in this habitat were *Gomphus externus*, *G. modestus*, and unidentified species of *Gomphus*, *Brechmorhaga mendax*, and *Tramea* prob. *onusta*. No mayflies that are carnivores were found in this habitat. The hemipteran *Ambrysus* sp., a carnivore, was collected. Also present in this lentic habitat were the carnivorous *Notonecta undulata* and *Saldula* sp. No beetles were found. In the soft mud were found the carnivorous Chironomids *Conchapelopia* sp., *Zavrelimyia* sp., and *Cryptochironomus fulvus*; and *Tabanus* sp.

The carnivorous eel *Anguilla rostrata* was found among the vegetation in this habitat. A carnivorous fish found in this lentic habitat was *Gila nigrescens*. Several carnivorous catfish were also present: *Ictalurus furcatus*, *I. melas*, and *I. natalis*. Another fish belonging to this family found in this habitat was the carnivorous *Pylodic-tis olivaris*. Several carnivorous centrarchids were found in these lentic habitats, particularly in pools but sometimes feeding along the margins of the stream: *Micropterus salmoides*, *Chaenobryttus gulosus*, *Lepomis cyanellus*, *L. gulosus*, *L. megalotis*, and *Pomoxis nigromaculatus*. Another carnivore found was *Perca flavescens*. The eel *Morone chrysops*, a carnivore, was also present in these lentic habitats, as was *Aplodinotus grunniens*.

Lotic Habitat.   In the areas where the current was relatively fast, the diversity was not as great as in the lentic habitats. Attached to rocks and rubble were the diatoms *Cocconeis pediculus* and *C. placentula*. Also attached to the rocks in areas where the current was not as fast was *Melosira granulata*. Another diatom present in this habitat was *Gomphonema olivaceum*. It was present where the current was not as fast, usually attached to rocks.

The sponge, *Dosilia palmeri*, an omnivore, was also found attached to the rocks. Where the current was not as fast were the omnivorous colonial hydrus, *Hydra* sp.; the molluscs *Ferrissia fragilis* and an unidentified species of the same genus. Both of these are omnivores. Crawling over the rocks where current was not as rapid were *Physa virgata*. This is an herbivore-omnivore. Embedded in the coarse gravel was the omnivorous unionid, *Anodonta*. Also in areas where the current was not as great was the omnivorous *Corbicula manilensis*. In among the rocks were two crayfish, *Orconectes causeyi* and *Procambarus clarkii*.

Dragonflies were also present in these lotic habitats. They were the carnivorous *Hetaerina* sp. and *Argia* sp. These were often among vegetation. Several gomphids were present, usually among the vegetation but sometimes in between the rocks. They were the carnivorous *Erpetogomphus* sp. and species of *Gomphus*, *G. externus*, *G. modestus*, and unidentified species of *Gomphus*. Other carnivores in among the rocks in fairly rapid current were *Ambrysus* sp., *Cryphocricos* sp., *Limnocoris* sp., *Notonecta undulata*, and *Rhagovelia choreutes*. Several carnivorous beetles were present. They were the omnivore-carnivore, *Dytiscus* sp., and the carnivorous *Laccophilus* sp. *Larsia* sp., a carnivore, was also present, as was the carnivorous *Zavrelimyia* sp.

In the lotic habitat among the rocks, particularly in riffles and in areas where the current was relatively fast, was found the eel, *Anguilla rostrata*, a carnivore. Also present in this habitat were *Hybognathus placita*, a carnivore, and *Rhinichthys cataractae*, a carnivore. A catfish, *Pylodictis olivaris*, was found in among the rocks. The fast-flowing water is an ideal habitat for many of the salmonid fishes. Found in this habitat were *Esox lucius*, *Salmo clarki*, *S. gairdneri*, and *S. trutta*, all carnivores. Other carnivores in this habitat were the centrarchids, *Micropterus dolomieui* and *M. salmoides*. *Chaenobrytus gulosus* and *Lepomis cyanellus*, both carnivores, were found in the fast-flowing water, as was *Lepomis macrochirus*, an omnivore. The carnivore, *Lepomis megalotis*, was also present. Of the Percidae, the carnivorous *Perca flavescens* and *Stizostedion vitreum* were also present. In among the rocks was the carnivorous *Morone chrysops*.

SUMMARY

The Rio Grande is the fifth-longest river in the United States. It is formed or "rises" in the mountains of southern Colorado, flows through New Mexico, forms the border between Mexico and the United States as it flows through Texas, and theoretically, empties into the Gulf of Mexico. Unfortunately, due to diversion in recent years, it no longer flows into the Gulf. Over its history, it has been dammed and diverted mainly for agricultural activities, but also for related activities. Some mining has been done in the watershed and the water has been diverted for this purpose. Groundwater, which normally would flow into the river, has also been diverted for irrigation. The Rio Grande is dammed several times: Heron Dam, El Vado, Abiquiu, McClure, Cochiti, Jemez Canyon, Galisteo, Bluewater, Caballo, American Diversion Dam and International Dam in El Paso, Amistad, Falcon, Anzalduas, and Retamal. Over the years the river has been subjected to several geological events, particularly uprisings.

The Rio Grande basin has an arid to semiarid climate. It has relatively low humidity, with wide variations in diurnal temperature. The drainage basin upstream from Bernalillo has a semiarid climate, whereas downstream from Belen the climate is arid. There is wide variation in precipitation at various times of the year. It is subject to frequent storms which arise mainly from the Gulf of Mexico. Rainfall is irregular and somewhat unpredictable. The period of transition of winter to summer, March through May and September through October, is associated with the largest storms. These generate floods.

The vegetation of the watershed is divided roughly into the Woodland Biome, found at elevations from 4500 to 7500 ft above sea level; the Mixed Grassland Association, which is a combination of both midheight grasses and short grasses; and the Desert Grassland Association. The Mixed Grassland Association is the Association that forms the immediate banks of the Rio Grande. It is found on the lower elevations and occupies a large part of the rolling plains that border the lower desert grasslands, through which the Rio Grande flows. Because of overgrazing and drought, this association has degraded, and shrubs such as snakeweed, rabbitbrush, shadscale, and saltbush are rapidly invading the area. The Desert Grassland Association extends across the southern part of New Mexico with northern extensions into the major river valleys. In the Rio Grande watershed of New Mexico, this grassland covers the southern portion of the watershed and extends up the middle of the Rio Grande valley almost as far north as the town of Bernalillo. The dominant plants in this association are creosote bush, tarbush, mesquite, and so on.

In its upper reaches, the river channel is a deep irregular canyon with rapids, and the river is very deep. In the upper reaches the river flow is very fast. Below the town of Embudo the river velocity slows down as it flows into the Española valley. In this reach the river is funneled through the irrigation cycle of diversion and return channels as it is used for agricultural purposes. After it leaves this valley, the Rio Grande increases in velocity for the next 25 mi as it passes through the gorge known as White Rock Canyon. This area marks the southern limit of the fairly fast-flowing water of the Rio Grande in New Mexico. It flows through a broad valley until it reaches Otowi Bridge (just south of Santa Clara Pueblo), where it becomes abruptly constricted as it enters another canyon that extends downstream to Cochiti Pueblo. The remainder of the river course through New Mexico is through a broad valley of low relief. Stream gradients of 4.75 m km^{-1} are common upstream from Cochiti Pueblo, while the gradient downstream is 0.95 m km^{-1} prevailing through Cochiti Pueblo and El Paso, Texas. In Texas, the gradient of the river is low. The sediment load of the Rio Grande is very large and is trapped behind dams in many areas.

Major tributaries of the Rio Grande are the Rio Chama, Jemez River, Rio Puerco, Rio Conchos, Pecos River, Devil's River, and Rio Salado. These tributaries bring large amounts of water to the Rio Grande and also bring in considerable amounts of sediment. For example, from the Rio Puerco 82% of the sediment transport may take place during storm events that reoccur about once a year. Streamflow regulation by dams has caused several physical components of the fluvial system to adjust to new sediment and water discharge conditions, including sediments delivered by tributaries. Tributaries often contribute high sediment loads, which has distinct effects on the channel pattern from straight to meandering to braided. The channel capacity may decrease and the floodplain area may aggrade and increase. Groundwater may also change, which supports large stands of phreatophytes. These transpire large amounts of water. As a result, dramatic losses in water delivery to the downstream reservoirs occur. Tectonics have also greatly influenced the riverine system.

The physical characteristics of the Rio Grande channel may have been profoundly affected by the construction of dams. For example, since the construction of the Cochiti Dam, there has been more consistency in flow of the Rio Grande and a reduction in the frequency of large seasonal floods. These dams have damped the long-term cycles in flow of the river and thus reduced flooding. Dams such as Elephant Butte Dam have had definite effects on the upstream flow of the river as well as downstream. The reservoir produces a new base level for upstream reaches of the river, and the upstream rise of the base level of the reservoir decreases the slope of the upstream reach and thus changes the fluvial geomorphology and hydraulic geometry as well as sediment transport. The creation of the reservoir behind a dam causes a decrease in river slope upstream, and this was accompanied by a decrease in flow velocity. As a result, there were depositions and formation of a large fine-grained delta. This resulted in the development of a system of anastomosing channels and the loss of a single, distinct channel. This has resulted in encroachment and establishment of thick stands of riparian vegetation.

The pollutants that were formerly transported by the Rio Grande are now being deposited in the reservoir. Examples of these are cadmium, mercury, lead, uranium, and pesticides. These substances are bound primarily to the sediments. An unknown amount is being cycled from the sediments to the water in the reservoir. The channel of the Rio Grande is very changeable depending on its location in the drainage basin and the proximity of tributaries that discharge a large volume of sediments into the main stem.

The flow of the Rio Grande is very variable and affected greatly by the positioning of dams in the river. It is also affected greatly by the entrance of tributaries, which sometimes have very large flows. A considerable amount of water is diverted into Mexico. The amount is roughly 65 million cubic meters annually. This is used for irrigation of the Juarez Valley. The Pecos and Devil's Rivers contribute roughly 274 to 353 million cubic meters annually. Below Laredo, the Rio Salado contributes 472 million cubic meters annually, and the combined flow reaches 3 billion cubic meters annually at Falcon International Reservoir. Below Falcon, the Rio San Juan contributes 434 million cubic meters a year. The Rio Grande is diverted between Rio Grande City and Anzalduas Dam at a rate of 292 million cubic meters a year for irrigation. The major diversion in Mexico is at Reynosa. On the U.S. side there is a combined diversion of 919 million cubic meters per year. When the Rio Grande reaches Brownsville, the flow decreases to 1.18 billion cubic meters per year, which includes erratic floodwaters after a storm. There is considerable groundwater recharge to the Rio Grande.

The chemical and physical characteristics of the water are greatly altered by inflows from various activities. In Water Year 1987–1988 the hardness of the Rio Grande varied from 220 to 410 mg L^{-1} as calcium carbonate. The noncarbonate hardness varied from 65 to 160 mg L^{-1}. Calcium was the main anion that contributed to its hardness. During water year 1987–1988 when these data were obtained, the chlorides varied from 77 to 200 mg L^{-1}.

Metals and other toxic materials have accumulated in Rio Grande sediments and, in some cases, in fish. Metals have been found in the water samples, particularly manganese, but also arsenic, barium, chromium, copper, lead, mercury, nickel, and zinc. Toxic metals have also been found in the stretch of the river from the river-side diversion dam in El Paso County to the International Dam in El Paso County. Two of the principal metals were lead and chromium. In concentrations slightly above the human health protection criteria, they were found at the Rio Conchos confluence with the Rio Grande in Presidio County (1979, 1994). They were also detected in samples taken between Laredo and Nuevo Laredo. In the 19 samples of dissolved lead collected between October 1981 and August 1986 there was a mean of 1.421 μg L^{-1} and a standard deviation of 1.042 μg L^{-1}.

Fecal coliforms were also found in the water. They were high from Laredo to Falcon Dam and Brownsville. Small amounts of dissolved chromium were found at the El Jardin Pumping Station, Brownsville. The highest reading of total chromium in this area was 30 μg L^{-1} which occurred on August 18, 1982. This was under the human health criteria for chromium. It was slightly above the health criteria at the confluence of the Rio Conchos. A high reading of 36.4 mg kg^{-1} occurred September 12, 1979 at Rio Conchos. Lead was found to be of slight to moderate potential for toxicity at the Rio Conchos confluence.

Despite the heavy pollution level, there is, at certain sections of the Rio Grande, considerable aquatic life. The functioning ecosystem can best be described in the reach of the Rio Grande above the American Dam (from the Caballo Dam to the American Dam). This is the area in which the most intensive study of the various kinds of aquatic life has been made and an ecosystem can be constructed.

BIBLIOGRAPHY

Academy of Natural Sciences of Philadelphia. 1975. Environmental assessment of the impact of PL 92-500 on the upper Rio Grande river. Final report submitted to the National Commission on Water Quality under contract WQ5ACO44. ANSP, Philadelphia, Pa.

Alves, J. 1998. Status of Rio Grande cutthroat trout in Colorado. Unpublished report of the Colorado Division of Wildlife, Denver, Colo. 10 pp.

American Rivers. 1993. Ten most endangered rivers for 1993. American Rivers, Washington, D.C. 6 pp.

Anderholm, S. K. 1997. Water-quality assessment of the Rio Grande valley, Colorado, New Mexico, and Texas: shallow ground-water quality and land use in the Albuquerque area, central New Mexico, 1993. Water-Resources Investigations Rep. 97-4067. U.S. Geological Survey, Washington, D.C. 73 pp.

Anderson, A. A., C. Hubbs, K. O. Winemiller, and R. J. Edwards. 1995. Texas freshwater fish assemblages following three decades of environmental change. Southwest. Nat. 40(3): 314–321.

Andreason, M. K. 1985. Insecticide resistance in mosquitofish of the lower Rio Grande valley of Texas: an ecological hazard? Arch. Environ. Contam. Toxicol. 14: 573–577.

Angermeier, P. L., and J. R. Karr. 1994. Biological integrity versus biological diversity as policy directives: protecting biotic resources. BioScience 44 (10): 690–697.

Arruda, J. A., M. S. Cringan, D. Gilliland, S. G. Haslouer, J. E. Fry, R. Broxterman, and K. L. Brunson. 1987. Correspondence between urban areas and the concentrations of chlordane in fish from the Kansas River. Bull. Environ. Contamin. Toxicol. 39: 563–570.

Baker, R. C., and O. C. Dale. 1964. Ground-water resources of the lower Rio Grande valley area, Texas. Water-Supply Pap. 1653. U.S. Geological Survey, Washington, D.C.

Baldwin, P. M. 1938. A short history of the Mesilla valley. N.M. Hist. Rev. 13: 314–324.

Behnke, R. J. 1972. The salmonid fishes of recently glaciated lakes. J. Fish. Res. Board Can. 29: 639–671.

Behnke, R. J. 1992. Native trout of western North America. Am. Fish. Soc. Monogr. 6: 1–275.

Bestgen, K. R., and S. P. Platania. 1990. Extirpation of *Notropis simus simus* (Cope) and *Notropis orca* Woolman (Pisces: Cyprinidae) from the Rio Grande in New Mexico, with notes on their life history. Occas. Pap. Mus. Southwest. Biol. 6: 1–8.

Big Bend Natural History Association. 1993. Big Bend National Park calendar. BBNHA, Big Bend National Park, Texas.

Bomar, G. W. 1983. Texas weather. Univ. Texas Press, Austin, Texas.

Bowles, W. F., Jr. 1983. Intensive survey of the Rio Grande near Laredo, segment 2304, bacteriological. Texas Department of Water Resources, Weslaco, Texas.

Brandvold, D. K., C. J. Popp, and J. A. Brierley. 1976. Waterfowl refuge effect on water quality: II. Chemical and physical parameters. J. Water Pollut. Control Fed. 48(4): 680–687.

Brierley, J. A., D. K. Brandvold, and C. J. Popp. 1975. Waterfowl refuge effect on water quality: I. Bacterial populations. J. Water Pollut. Control Fed. 47(7): 1892–1900.

Briones, P. O. 1994. Survey of heavy metals (chromium and lead) in water, mud, and fish deposits in the Rio Grande from El Paso to Brownsville. M.S. thesis. Univ. Texas of the Permian Basin, Odessa, Texas. 64 pp.

Bristow, B. A. 1993. Net winter productivity and trophic relationships in the Rio Grande between Elephant Butte Dam and Caballo Lake, New Mexico. M.S. thesis. New Mexico State Univ., Las Cruces, N. Mex.

Broderick, S. C. 2000. Fish and herpetological communities in wetlands in the Fort Craig reach of the Rio Grande and the Elephant Butte Delta, Socorro County, New Mexico 1997 to 1998. U.S. Bureau of Reclamation, Denver, Colo.

Brown, W. H. 1953. Introduced fish species in the Guadalupe River basin. Tex. J. Sci. 5: 245–251.

Bryan, K. 1928. Historic evidence on changes in the channel of Rio Puerco, a tributary of the Rio Grande in New Mexico. J. Geol. 36: 265–282.

Buckner, H. D., E. R. Carrillo, H. J. Davidson, and W. J. Shelby. 1988. Water resources data, Texas water year 1988. Water-Data Rep. TX-88-3. U.S. Geological Survey, Washington, D.C.

Bullard, T. F., and S. G. Wells. 1992. Hydrology of the middle Rio Grande from Velarde to Elephant Butte Reservoir, New Mexico. Resource Publ. 179. Fish and Wildlife Service, U.S. Department of the Interior, Washington, D.C.

Bureau of Reclamation, Albuquerque Area Office. 2000. Rio Grande and low flow conveyance channel modifications. Draft of environmental impact statement. U.S. Department of the Interior, Albuquerque, N. Mex.

Buzan, D. L. 1990. Intensive survey of the Rio Grande Segment 2304. March 14–17, 1988 and June 17–21, 1988. Texas Water Commission, Austin, Texas.

Calamusso, B., and J. N. Rinne. 1996. Distribution of the Rio Grande cutthroat trout and its co-occurrence with the Rio Grande sucker and Rio Grande chub on the Carson and Santa Fe national forests. Pp. 157–167. *In:* D. W. Shaw and D. M. Finch (tech. coords.), Desired future conditions for southwestern riparian ecosystems: bringing interests and concerns together. Sept. 18–22, 1995, Albuquerque, N. Mex. Gen. Tech. Rep. RM-GTR-272. U.S. Department of Agriculture, Forest Service, Rocky Mountain Forest and Range Experiment Station, Fort Collins, Colo. 359 pp.

Carter, L. F. 1997a. Water-quality assessment of the Rio Grande valley, Colorado, New Mexico, and Texas: organic compounds and trace elements in bed sediment and fish tissue, 1992–93. Water-Resour. Invest. Rep. 97-4002. National Water-Quality Assessment Program, U.S. Geological Survey, Albuquerque, N. Mex.

Carter, L. F. 1997b. Water-quality assessment of the Rio Grande valley, Colorado, New Mexico, and Texas: fish communities at selected sites, 1993–95. Water-Resour. Invest. Rep. 97-4017. U.S. Geological Survey, Washington, D.C.

Chaney, A. H., and M. Pons. 1989. Faunal and floral characteristics of the area to be affected by the Playa del Rio project site, Cameron County, TX. U.S. Fish and Wildlife Project Rep. 14-16-002-86-926. Ecological Branch, Corpus Christi, 12 pp.

Clark, I. 1987. Water in New Mexico: a history of its management and use. Univ. New Mexico Press, Albuquerque, N. Mex. Pp. 41–53.

Cope, E. D., and H. C. Yarrow. 1875. Report upon the collections of fishes made in portions of Nevada, Utah, California, Colorado, New Mexico, and Arizona during the years 1871–1874. Chap. 6, pp. 635–703. *In:* U.S. Army Engineers Department Report, in charge of George M. Wheeler, Geog. Geol. Expl. Surv. West of 100th Meridian 5: 1–1021.

Cordell, C., C. L. Long, and D. W. Jones. 1985. Geological expression of the batholith beneath Questa caldera, New Mexico. J. Geophys. Res. 90: 263–269.

Costa, J. E. 1984. Physical geomorphology of debris flows. Pp. 268–317. *In:* J. E. Costa and P. J. Fleisher (eds.), Developments and applications of geomorphology. Springer-Verlag, Berlin.

Dallas Morning News. 1994. Starr among poorest in U.S. February 8.

Davis, J. R. 1980. Species composition and diversity of benthic macroinvertebrates in the upper Rio Grande, Texas. Southwest Nat. 25(2): 137–150.

Davis, M. E., and E. R. Leggat. 1965. Reconnaissance investigation of the ground-water resources of the upper Rio Grande basin, Texas. Texas Water Commission, Austin, Texas.

Edwards, R. J., and S. Contreras-Balderas. 1991. Historical changes in the ichthyofauna of the lower Rio Grande (Rio Bravo del Norte), Texas and Mexico. Southwest. Nat. 36(2): 201–212.

Ellis, M. M. 1914. Fishes of Colorado. Univ. Colo. Stud. 11(1): 5–135.

Finch, D. M., J. C. Whitney, J. F. Kelly, and S. R. Loftin. 1998. Rio Grande ecosystems: linking land, water, and people. Toward a sustainable future for the middle Rio Grande basin. June 2–5. Albuquerque, N. Mex. Rocky Mountain Research Station, Ogden, Utah.

Flock, L. R. 1931. Effect of the operation of Elephant Butte Reservoir on the river through the Rio Grande Project. Unpublished report on file in the office of the International Boundary and Water Commission, El Paso, Texas.

Ford, D. A. 1996. An environmental evaluation of the natural springs and spring-fed creeks along the Rio Grande in the Big Bend area of Texas. M.S. thesis. Sul Ross State University, Alpine, Texas. 97 pp.

Gloyd, L. K. 1958. The dragonfly fauna of the Big Bend region of Trans-Pecos Texas. Occas. Pap. Mus. Zool., Univ. Mich. 593: 1–23 and several plates.

Goddard, J. A., and L. C. Redmond. 1978. Northern pike, tiger muskellunge, and walleye populations in Stockton Lake, Missouri: a management evaluation. Am. Fish. Spec. Publ. 11: 313–319.

Goodnight, C. J., and L. S. Whitley. 1960. Oligochaetes as indicators of pollution. Proc. 15th Industrial Waste Conference, Purdue Univ. Eng. Bull. 106: 139–142.

Graf, W. L. 1982. Spatial variation of fluvial processes in semiarid lands. Pp. 193–277. *In:* C. E. Thorn (ed.), Space and time in geomorphology. Allen & Unwin, New York.

Gregory, S. V., F. J. Swanson, W. A. McKee, and K. W. Cummins. 1991. An ecosystem perspective of riparian zones. BioScience 41: 540–550.

Hansmann, E. W., and N. J. Scott, Jr. 1977. A natural history survey of the Rio Grande valley between Bernarillo and Elephant Butte Reservoir. National Fish and Wildlife Laboratory, Museum of Southwestern Biology, Univ. New Mexico, Albuquerque, N. Mex.

Harman, W. J., M. S. Loden, and J. R. Davis. 1979. Aquatic Oligochaeta new to North America with some further records of species from Texas. Southwest. Nat. 24(3): 509–525.

Harrell, H. L. 1978. Response of the Devil's River (Texas) fish community to flooding. Copeia 1978: 60–68.

Hendrickson, J. A., Jr., L. F. Berseth, T. E. Walton III, and J. W. Richardson, Jr. 1975. Environmental assessment of the impact of PL 92-500 on the upper Rio Grande river. Final report submitted to the National Commission on Water Quality under contract WQ5ACO44. Academy of Natural Sciences of Philadelphia, Philadelphia, Pa.

Hill, M. T., W. S. Platts, and R. L. Beschta. 1991. Ecological and geomorphological concepts for instream and out-of-channel flow requirements. Rivers 2: 198–210.

Hodge, L. D. 1994. Poisoning a region's lifeblood. *In:* Texas parks and wildlife: On the border special issue. Texas Parks and Wildlife Department, Austin, Texas.

Howard, C. S., and S. K. Love. 1945. Quality of surface waters of the United States 1943 with a summary of analyses of streams in Colorado River, Pecos River, and Rio Grande basins, 1925 to 1943. Water-Supply Pap. 970. U.S. Geological Survey, Washington, D.C.

Hubbs, C., R. J. Edwards, and G. P. Garrett. 1991. An annotated checklist of the freshwater fishes of Texas, with keys to identification of species. Tex. J. Sci., Suppl. 43(4): 56 pp.

International Boundary and Water Commission (U.S. and Mexico). 1932. Report on the changes in the regime of the Rio Grande in the valleys below since the construction of Elephant Butte Dam, 1917–1932. Unpublished report on file in the office of the International Boundary and Water Commission, El Paso, Texas.

Jacobi, G. Z. 1980. Benthological monitoring at ambient stream stations on the Pecos, Rio Grande, and San Juan rivers, New Mexico, in 1979. Environmental Protection Agency, New Mexico State Biological Water Monitoring Program, Albuquerque, N. Mex.

Jahrsdoerfer, S. E., and D. M. Leslie, Jr. 1988. Tamaulipan brushland of the lower Rio Grande valley of south Texas: description, human impacts, and management options. Biol. Rep. 88(36). Fish and Wildlife Service, U.S. Department of the Interior, Washington, D.C.

Jenkins, D. W. 1981. Biological monitoring of toxic trace elements. Rep. 600/S3-80-090. U.S. Environmental Protection Agency, Washington, D.C. Pp. 1–9.

Jester, D. B. 1967. A new crayfish of the genus *Orconectes* from New Mexico (Decapoda, Astacidae). Am. Midl. Nat. 77(2): 518–524.

Jetton, E. V., and J. W. Kirby. 1970. A study of precipitation, streamflow, and water usage on the upper Rio Grande. Atmos. Sci. Group Rep. 25. Univ. Texas, Austin, Texas.

Jordan, D. S., and B. W. Evermann. 1896. The fishes of North and Middle America. *Bull. U.S. Nat. Mus.*, 47: 1–3313.

Judd, F. W. 1994. Report of literature review on discharges from the Rio Grande and Arroyo Colorado and their impacts (draft). University of Texas–Pan American Coastal Studies Laboratory, Edinburg, Texas.

Kelsch, S. W., and F. S. Hendricks. 1990. Distribution of the headwater catfish *Ictalurus lupus* (Osteichthyes: Ictaluridae). Southwest. Nat. 35(3): 292–297.

Kirkham, P. M., and J. D. Holm. 1988. Environmental problems and reclamation activities at inactive metal mines and milling sites in San Luis valley, Colorado. Mined Land Reclamation Division and Colorado Department of Health, Water Quality Control Division, Denver.

Koster, W. J. 1957. Guide to the fishes of New Mexico. Univ. New Mexico Press, Albuquerque, N. Mex.

Lagasse, P. F. 1980. An assessment of the response of the Rio Grande to dam construction: Cochiti to Isleta reach. Tech. Rep. U.S. Army Corps of Engineers, Albuquerque District, Albuquerque, N. Mex. 133 pp. + append.

Lagasse, P. F. 1981. Geomorphic response of the Rio Grande to dam construction. Pp. 27–46. *In:* S. G. Wells and W. Lambert (eds.), Environmental geology and hydrology in New Mexico. Spec. publ. 10. New Mexico Geological Society, Albuquerque, N. Mex.

Langlois, D., J. Alves, and J. Apker. 1994. Rio Grande sucker recovery plan. Colorado Division of Wildlife, Denver. 22 pp.

Lee, W. T. 1907. Water resources of the Rio Grande valley in New Mexico. Water-Supply Irrig. Pap. 188. U.S. Geological Survey, Washington, D.C. Pp. 1–59.

Leopold, L. B., M. G. Wolman, and J. P. Miller. 1964. Fluvial processes in geomorphology. W. H. Freeman, San Francisco. 522 pp.

Levings, G. W., D. F. Healy, S. F. Richey, and L. F. Carter. 1998. Water quality in the Rio Grande valley, Colorado, New Mexico, and Texas, 1992–95. Circ. 1162. U.S. Geological Survey, Washington, D.C.

Lind, O. T., and C. A. Bane. 1975. A limnological survey of aquatic resources of the Rio Grande valley in Big Bend National Park between Hot Springs and Boquillas Canyon. Project order PX700-3-0502. Prepared for the Office of Natural Science, Southwest Region National Park Service. Big Bend National Park.

Lubinski, K. S., A. Van Vooren, G. Farabee, J. Janeck, and S. D. Jackson. 1986. Common carp in the upper Mississippi River. Hydrobiologia 136: 141–154.

Luecke, C. 1986. Ontogenetic changes in feeding habits of juvenile cutthroat trout. Trans. Am. Fish. Soc. 115: 703–710.

Maker, H. J., J. M. Downs, and J. U. Anderson. 1972. Soil associations and land classification for irrigation, Sierra County. Agric. Exp. Res. St. Rep. 233. New Mexico State Univ., Las Cruces, N. Mex.

Maret, T. R., and E. J. Peters. 1979. Food habits of the white crappie, *Pomoxis annularis* Rafinesque, in Branched Oak Lake, Nebraska. Trans. Nebr. Acad. Sci. VII: 75–82.

Maxwell, R. A. 1990. The Big Bend of the Rio Grande. Bureau of Economic Geology, Univ. Texas, Austin, Texas. Pp. 114–119.

McGowan, N., R. J. Kemp, Jr., and R. McCune. 1971. Freshwater fishes of Texas. Bull. 5-A. Texas Parks and Wildlife Department. Austin, Texas.

McLemore, D. 1994. Dangerous waters. *Dallas Morning News*. Pp. 47A–57A.

Mendieta, H. B. 1974. Reconnaissance of the chemical quality of surface waters of the Rio Grande basin, Texas. Texas Water Dev. Board Rep. 180. Prepared by U.S. Geological Survey under cooperative agreement with the Texas Water Development Board.

Metcalf, A. L. 1967. Later Quaternary mollusks of the Rio Grande valley, Caballo Dam, New Mexico to El Paso, Texas. Texas Western Press, Univ. Texas, El Paso, Texas. Sci. Ser. 1: 1–62.

Miller, R. R. 1961. Man and the changing fish fauna of the American southwest. Mich. Acad. Sci. Arts Lett. 46: 365–404.

Minckley, W. L. 1973. Fishes of Arizona. Arizona Game and Fish Department, Phoenix, Ariz.

Minckley, W. L. 1980. *Catostomus plebius* Baird and Girard, Rio Grande sucker. Pp. 387. *In:* D. C. Lee et al., (eds.), Atlas of North American freshwater fishes. North Carolina State Museum of Natural History, Raleigh, N.C.

New Mexico Environmental Improvement Agency, Water Quality Division. 1973. Upper Rio Grande Basin Plan. NMEIA, Santa Fe, N. Mex.

New Mexico Geological Society. 1971. *In:* H. L. James (ed.), Guidebook to the San Luis basin, Colorado: 22nd Field Conference. NMGS, Santa Fe, N. Mex.

New Mexico Geological Society. 1981. New Mexico highway geological map: Socorro, New Mexico. Bureau of Mines and Mineral Resources, NMGS, Santa Fe, N. Mex.

New Mexico State Engineer. 1967. Water resources of New Mexico: Occurrence, development, and use. State Planning Office, Santa Fe, N. Mex.

New Mexico Water Quality Control Commission. 1975. Water quality in New Mexico. NMWQCC, Santa Fe, N. Mex.

New Mexico Water Quality Control Commission. 1976. Upper Rio Grande Basin Plan. NMWQCC, Santa Fe, N. Mex.

New Mexico Water Resources Research Institute. 1973. An analytical interdisciplinary evaluation of the utilization of the water resources of the Rio Grande in New Mexico. Technical completion report projects B-026, B-019, and B-016-NMEX. NMWRRI, New Mexico State Univ., Las Cruces, N. Mex.

New Mexico Water Resources Research Institute. 1994. The water future of Albuquerque and the middle Rio Grande: Proc. 39th Annual New Mexico Water Conf. Nov. 3–4, Albuquerque Convention Center.

Nordin, C. F., Jr. 1963. A preliminary study of sediment transport parameters, Rio Puerco near Bernardo, New Mexico. Prof. Pap. 462-C. U.S. Geological Survey, Washington, D.C.

Nordin, C. F., Jr. 1964. Aspects of flow resistance and sediment transport Rio Grande near Bernalillo, New Mexico: studies of flow in alluvial channels. Water-Supply Pap. 1498-H. U.S. Geological Survey, Washington, D.C.

Ottmers, D. 1979. Intensive surface water monitoring survey for segment no. 2308 Rio Grande (Riverside Diversion Dam to New Mexico state line). IMS-82. Texas Department of Water Resources, Austin, Texas.

Ouchi, S. 1983. Effects of uplift on the Rio Grande over the Socorro magma body, New Mexico. Pp. 54–56. *In:* C. E. Chapin (ed.), Socorro Region II. Guidebook 34. New Mexico Geological Society, Santa Fe, N. Mex.

Ouchi, S. 1985. Response of alluvial rivers to slow active tectonic movement. Geol. Soc. Am. Bull. 96: 504–515.

Ozmina, D. J. 1965. An evaluation of the physical, chemical, and biological factors of Caballo Reservoir following rehabilitation. M.S. thesis. New Mexico State Univ., Las Cruces, N. Mex.

Peter, K. D. 1987. Ground-water flow and shallow-aquifer properties in the Rio Grande inner valley south of Albuquerque, Bernalillo County, New Mexico. Water-Resources Investigation Rep. 87-4015. U.S. Geological Survey, Washington, D.C. 29 pp.

Poff, N. J., and J. V. Ward. 1989. Implications of streamflow variability and predictability for lotic community structure; a regional analysis of streamflow patterns. Can. J. Fish. Aquat. Sci. 46: 1805–1818.

Popp, C. J. 1978. Heavy metals in water and selected tissue samples of aquatic life in the middle Rio Grande valley in New Mexico. Interim report to New Mexico Environmental Improvement Agency.

Popp, C. J., D. K. Brandvold, T. R. Lynch, and L. A. Brandvold. 1983. An evaluation of sediments in the middle Rio Grande, Elephant Butte Reservoir and Caballo Reservoir as potential sources for toxic materials. Technical completion report, project 1412626, to New Mexico Water Resources Research Institute.

Rathbun, R. E., V. C. Kennedy, and J. K. Culbertson. 1971. Transport and dispersion of fluorescent tracer particles for the flat-bed condition, Rio Grande Conveyance Channel, near Bernardo, New Mexico. Prof. Pap. 562-I. U.S. Geological Survey, Washington, D.C.

Reynolds, R. 1975. Rio Grande. C.H. Belding Graphic Arts Center Publishing, Portland, Oreg. 127 pp.

Rivera, J. A. 1998. Water democracies on the upper Rio Grande, 1598–1998. Pp. 20–27. *In:* D. M. Finch et al. (eds.), Rio Grande ecosystems: linking land, water, and people. Toward a sustainable future for the middle Rio Grande basin. June 2–5. Albuquerque, N. Mex. Rocky Mountain Research Station, Ogden, Utah.

Rocky Mountain Research Station. 1998. Rio Grande ecosystems: linking land, water, and people. Toward a sustainable future for the middle Rio Grande basin. June 2–5. Albuquerque, N. Mex. Rocky Mountain Research Station. Ogden, Utah.

Rosgen, D. L. 1994. A classification of natural rivers. Catena 22: 164–199.

Roy, R., T. F. O'Brien, and M. Rusk-Maghini. 1992. Organochlorine and trace element contaminant investigation of the Rio Grande, New Mexico. U.S. Fish and Wildlife Service, New Mexico Ecological Services Office, Albuquerque, N. Mex. 39 pp.

Runyon, R. 1947. Vernacular names of plants indigenous to the lower Rio Grande valley of Texas. Brownsville News Publishing, Brownsville, Texas.

Salomons, W., and U. Forstner. 1984. Metals in the hydrocycle. Springer-Verlag, Berlin.

Sanchez, C., Jr. 1970. Life history and ecology of carp, *Cyprinus carpio* Linnaeus, in Elephant Butte Lake, New Mexico. M.S. thesis. New Mexico State Univ., Las Cruces, N. Mex.

Schmitt, C. J., and W. G. Brumbaugh. 1990. National Contaminant Biomonitoring Program. Concentrations of arsenic, cadmium, copper, lead, mercury, selenium, and zinc in U.S. freshwater fish, 1976–1984. Arch. Contam. Toxicol. 19: 731–747.

Schmitt, C. J., J. L. Zajicek, and M. A. Ribick. 1985. National Contaminant Biomonitoring Program. Residues of organochlorine chemicals in U.S. freshwater fish, 1976–1984. Arch. Environ. Contamin. Toxicol. 19: 748–781.

Schumm, S. A. 1977. The fluvial system. Wiley, New York. 338 pp.

Scofield, C. S. 1938. Quality of water of the Rio Grande basin above Fort Quitman, TX. Water-Supply Pap. 839. U.S. Geological Survey, Washington, D.C.

Shacklette, H. T., and J. G. Boerngen. 1984. Element concentrations in soils and other surficial materials of the conterminous United States. Prof. Pap. 1270. U.S. Geological Survey, Washington, D.C. 105 pp.

Shupe, S. J., and J. Folk-Williams. 1988. The upper Rio Grande: a guide to decision-making. Western Network. Santa Fe, New Mexico.

Smith, G. R. 1966. Distribution and evolution of the North American catostomid fishes of the subgenus *Pantosteus*, genus *Catostomus*. Univ. Mich. Mus. Zool. Misc. Publ. 129. 132 pp.

Sodrensen, E. F., and D. Linford. 1967. Rio Grande basin: settlement, development, and water use. Pp. 143–168. *In:* Water resources of New Mexico: occurrence, development, and use. New Mexico State Engineer, State Planning Office, Santa Fe, N. Mex.

Stromberg, J. C. 1993. Instream flow models for mixed deciduous riparian vegetation within a semiarid region. Regul. Rivers Res. Manage. 8: 225–235.

Stromberg, J. C., and D. T. Patten. 1990. Riparian vegetation instream flow requirements: a case study from a diverted stream in the eastern Sierra Nevada, California. Environ. Manage. 14: 185–194.

Stromberg, J. C., D. T. Patten, and B. D. Richter. 1991. Flood flows and dynamics of Sonoran riparian forests. Rivers 2(3): 221–235.

Stumpff, W. K., and J. Cooper. 1996. Rio Grande cutthroat trout, *Oncorhynchus clarki virginalis*. *In:* D. E. Duff (ed.), Conservation assessment for inland cutthroat trout. USDA Forest Service, Intermountain Region, Ogden, Utah.

Sublette, J. E., M. D. Hatch, and M. Sublette. 1990. The fishes of New Mexico. Univ. New Mexico Press, Albuquerque, N. Mex.

Tamayo, J. L., and R. C. West. 1964. The hydrography of Middle America. *In:* Handbook of Middle American Indians. Univ. Texas Press, Austin, Tex.

Texas Natural Resources Conservation Commission. 1994. Regional assessment of water quality in the Rio Grande basin, including the Pecos River, the Devil's River, the Arroyo Colorado, and the lower Laguna Madre. Watershed Management Division, TNRCC, Austin, Texas.

Texas Natural Resources Conservation Commission. 1996. 1996 Regional assessment of water quality in the Rio Grande basin, including the Pecos River, the Devil's River, the Arroyo Colorado, and the lower Laguna Madre. SFR-041. Border Environmental Assessment Team, Office of Water Resource Management. TNRCC, Austin, Texas.

Texas Natural Resources Conservation Commission. 1998a. The Rio Grande Valley's future: nonpoint source and water quality seminar. Mar. 9–14, Brownsville, Texas.

Texas Natural Resources Conservation Commission. 1998b. Waste load evaluation for dissolved oxygen in the Rio Grande near El Paso in the Rio Grande basin segment 2308 and segment 2314. AS173. Water Quality Division, TNRCC, Austin, Texas.

Texas Water Commission. 1992a. Regional assessment of water quality in the Rio Grande basin. Pp. 71–196. Rep. GP 92-02. TWC, Austin, Texas.

Texas Water Commission. 1992b. The state of Texas water quality inventory, 11th ed. Rep. LP 92-16. TWC, Austin, Texas. 681 pp.

Tinkham, E. R. 1934. The dragonfly fauna of Presidio and Jeff Davis counties of the Big Bend region of Trans-Pecos, Texas. Can. Entomol. 66(10): 213–218.

Trowbridge, A. C. 1923. A geologic reconnaissance in the Gulf Coastal Plain of Texas, near the Rio Grande. Prof. Pap. 131-D. U.S. Geological Survey, Washington, D.C.

Tuan, Yi-Fu, C. E. Everard, and J. G. Widdison. 1973. The climate of New Mexico. State Planning Office, Santa Fe, N. Mex.

U.S. Environmental Protection Agency. 1971. A water quality survey of Red River and Rio Grande, New Mexico. USEPA, Washington, D.C.

U.S. Food and Drug Administration. 1984. Shellfish sanitation interpretation: action levels for chemical and poisonous substances. HFF-344. Food and Drug Administration, Center for Food Safety and Applied Nutrition, Shellfish Sanitation Branch, Washington, D.C. 4 pp.

U.S. Geological Survey. 1987. Water resources data, New Mexico, water year 1987. Water-Data Rep. NM-87-1. U.S. Geological Survey, Washington, D.C. 450 pp.

van der Leeden, F., F. L. Troise, and D. K. Todd. 1990. Longest rivers in the world. *In:* The water encyclopedia. Lewis Publishers, Chelsea, Mich.

Velz, E. J. 1984. Applied stream sanitation, 2nd ed. Wiley, New York.

West, S. W., and W. L. Broadhurst. 1975. Summary appraisals of the Nation's ground-water resources: Rio Grande region. Prof. Pap. 813-D. U.S. Geological Survey, Washington, D.C.

White, D. H., C. A. Mitchell, H. D. Kennedy, A. J. Krynitsky, and M. A. Ribick. 1983. Elevated DDE and toxaphene residues in fishes and birds reflects local contamination in the lower Rio Grande valley, Texas. Southwest. Nat. 28: 325–333.

Whitney, J. C. 1998. Watershed/river channel linkages: the upper Rio Grande basin and the middle Rio Grande bosque. Pp. 45–51. *In:* D. M. Finch et al. (eds.), Rio Grande ecosystems: linking land, water, and people. Toward a sustainable future for the middle Rio Grande basin. June 2–5, Albuquerque, N. Mex. Rocky Mountain Research Station, Ogden, Utah.

Wiener, J. G., G. A. Jackson, T. W. May, and B. P. Cole. 1984. Longitudinal distribution of trace elements (As, Cd, Cr, Hg, Pb, and Se) in fishes and sediments in the upper Mississippi River. Pp. 139–170. *In:* J. G. Wiener, R. V. Anderson, and D. R. McConville (eds.), Contaminants of the upper Mississippi River. Butterworth, Stoneham, Mass.

Wilcox, R. 1997. Concentrations of selected trace elements and other constituents in the Rio Grande and in fish tissue in the vicinity of Albuquerque, New Mexico, 1994 to 1996. Open File Rep. 97-667. U.S. Geological Survey, Washington, D.C.

Wilhm, J. L. 1967. Comparison of some diversity indices applied to populations of benthic macroinvertebrates in a stream receiving organic wastes. J. Water Pollut. Control Fed. 39: 1673–1683.

Wilson, D. C., and C. E. Bond. 1969. Effects of the herbicides diquat and dichlobenil (Casoron) on pond invertebrates: I. Acute toxicity. Trans. Am. Fish. Soc. 98: 438–443.

Wood, M. G. 1986. Life history characteristics of introduced blue tilapia, *Oreochromis aureus*, in the lower Rio Grande. M.S. thesis, Pan American Univ., Edinburg, Texas. 75 pp.

Wozniak, F. E. 1987. Irrigation in the Rio Grande valley, New Mexico: a study of the development of irrigation systems before 1945. Contract BOR-87-1. New Mexico Historic Preservation Division, Santa Fe, and U.S. Bureau of Reclamation, Southwest Regional Office, Amarillo, Texas. 191 pp.

Zimmerman, J. R. 1971. A comparison of five watersheds of New Mexico as interpreted from the distribution of the species of the aquatic beetle family Dytiscidae. Completion report, Water Resources Research Project 3109-21-A-010-NMEX.

Zuckerman, L. D., and D. Langlois. 1990. Status of Rio Grande sucker and chub in Colorado. Unpublished report. Colorado Division of Wildlife, Montrose, Colo. 44 pp.

CHAPTER 4

Pecos Riverine System

INTRODUCTION

River Course

The Pecos, a tributary of the Rio Grande, originates in the mountains of the Sangre de Cristo Range (Figure 4.1) where it is fed by snowmelt and rainfall. From its headwaters in the glaciated Sangre de Cristo Mountains in northern New Mexico, the Pecos flows 1320 km southeastward until it converges with the Rio Grande. The river flows southward through eastern New Mexico 700 km before it enters Texas. Throughout its passage flow is augmented by scant tributaries and springs and reduced by irrigation withdrawal, losses, and evaporation. The New Mexico watershed of the Pecos is approximately 25,992 mi^2 [New Mexico Water Quality Control Commission (NMWQCC), 1976]. The Pecos enters Texas approximately 25 km north of Orla. The length of the Pecos from New Mexico to where it joins the Rio Grande is about 620 km.

The Pecos River is characterized from its origin to Ft. Sumner, New Mexico, as a clear, freshwater mountain stream. This was caused by the influence of snowmelt, rainfall, and groundwater inflow. As the river approaches Roswell, New Mexico it begins to dissolve Permian salts. These contribute to the saline nature of the river. At Malaga, additional salts are added to the river. These salts are from saline springs. At Malaga, where the brine enters the main stem, it contributes about 109 million

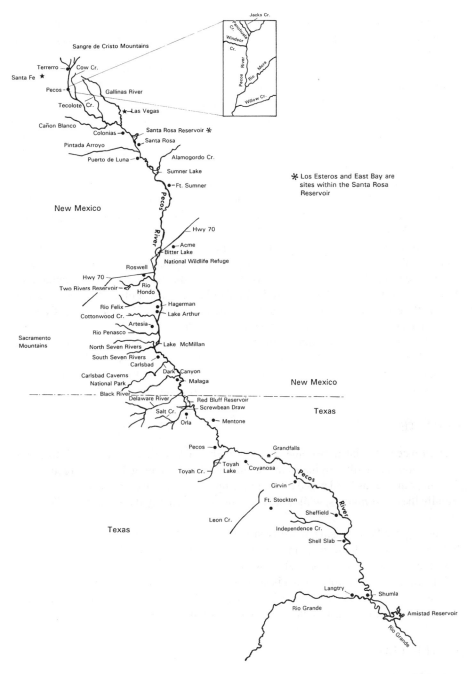

FIGURE 4.1. Map of Pecos River. (Created by Susan Durdu; drawn by Su-Ing Yong.)

kilograms (240 million pounds) of salt per year (Davis, 1987). The rate of flow is about 0.14 m^3 s^{-1} (4.94 ft^3/s) [Texas Natural Resources Conservation Commission (TNRCC), 1996b).

Once the Pecos enters Texas it flows into the Red Bluff Reservoir, a Public Works Administration project that has a capacity of about 310,000 acre-ft. It was completed in 1936 for flood control in order to equalize the flow for irrigation needs in the Trans-Pecos Region (Bailey, 1974). The river increases in salinity as it meanders from Red Bluff Reservoir to Girvin. This trend is attributed to continued dissolution of Permian salts and to the high evaporation rates due to the arid climate (Larkin and Bomar, 1983; TNRCC, 1996b).

The river flows out of Permian formations into Cretaceous limestone of the Edwards–Stockton Plateau as it reaches Sheffield. Freshwater flow is increased in this area from groundwater, springs, and tributaries. This brings about a dilution in the salts in the river (Davis, 1987; TNRCC, 1996b).

Tributaries. Major tributaries in the upper river include Cow Creek, Rio Mora, Gallinas River, Tecolote Creek, Cañon Blanco, Pintada Arroyo, and Alamogordo Creek. Between Sumner Lake and Carlsbad most tributaries arise in the Sacramento Mountains to the west. These include the Rio Hondo, Rio Felix, Rio Penasco, and Seven Rivers. In the Carlsbad area, the principal drainages include Dark Canyon, Black River, and Delaware River (NMWQCC, 1976).

The Pecos River is regulated by Sumner Lake, Two Rivers Reservoir, Lake McMillan, and Red Bluff Reservoir. These impoundments are used for flood control, irrigation, recreation, and sediment retention. Numerous small impoundments are located elsewhere. These are primarily to control sediment and runoff.

CLIMATE

The climate of the basin becomes progressively more xeric from the north to south. Precipitation typically originates in the Gulf of Mexico. About 12 in. of precipitation falls near the Texas border and 35 in. in the upper river. November and February are usually the driest months, while July and August are usually the wettest (NMWQCC, 1976).

Through the Pecos basin and westward to Big Bend (on the Rio Grande), mean rainfall diminishes rapidly from 20 in. to 8 in. Rainfall increases greatly in the Trans-Pecos with increasing altitude. Upper elevations may receive as much as 20 in./yr. Precipitation within the Edwards Plateau is generally in the range of 14 to 16 in./yr. This decreases to less than 12 in. in the Pecos valley. The mountains of the Sacramento section receive more than 14 in. and possibly as much as 20 in. (Mendieta, 1974).

WATERSHED

New Mexico
The New Mexican watershed of the Pecos River is approximately 25,992 mi^2. Aquifers in the basin are primarily unconsolidated sand and gravel, limestone, and sandstone. As early as 1942, the basin had been characterized as representing the

greatest aggregation of problems of land and water use of any irrigated basin in the western United States. These involved salinity, erosion, reservoir siltation, flooding, and interstate controversy concerning water use.

Flooding in the basin results from surface runoff following frontal or tropical storms (snowmelt is not a major source of flooding). Floods are short and typically occur in late summer months. In general, acquifers in the Pecos basin are hydraulically connected to the river (NMWQCC, 1976).

Although surface water flows are largely undependable, a few perennial reaches of Pecos are dependable because of substantial groundwater seepage. One seepage area lies 16 km south of Fort Sumner and extends for 56 km. Groundwater inflow of at least 0.28 to 0.85 mg s^{-1} is common throughout the year, but this is dissipated by the time the river reaches Highway 70. The other seepage area is between Hagerman and Artesia. Groundwater seepage offers flow stability in an otherwise hostile and unpredictable environment. Successful reproduction and survival of fish and other organisms appear to be restricted to these perennial seepage zones (Hatch et al., 1985).

Texas
The Pecos River in western Texas is in an arid region with an average annual rainfall that increases downstream from 30 cm at Orla to 41 cm at the mouth. Elevation above mean sea level decreases from 832 m near Orla to 345 m near the mouth, which is about 602 km downstream (Davis, 1980).

The river below Toyah Lake (upper Pecos in Texas) is deeply incised in the Comanchean Cretaceous limestone of the Stockton Plateau, the Trans-Pecos extension of the Edwards Plateau (Spiers and Hejl, 1970). Steep limestone bluffs occur on one or both banks in the reach.

Riverside Vegetation. The river banks in the reach from Red Bluff Reservoir to a point about 35 km south of Girvin support dense growths of salt cedar (*Tamarix* spp.). Extensive growths of the cane grass *Phragmites* occur near the mouth. Species found in the watershed are listed in Table 4.6.

GEOLOGY

The early Tertiary uplifting of the San Juan Mountains in southwest Colorado had created the ancestral Rio Grande–Pecos system that may have either flowed into the lower Pecos, the Colorado River, or to the western Gulf. An example of this is the Brazos River. Meiocene uplifts of the Sangre de Cristo Mountains in New Mexico beheaded the drainage and separated the headwaters of the Rio Grande from those of the Pecos (Echelle and Echelle, 1978).

Most of the geology of the basin is dominated by sedimentary deposits. Igneous rocks are most plentiful in the western basin, particularly in the mountains. Most of the water-bearing rocks are located in the plains. Water-bearing deposits of limestone, anhydrite, and halite are plentiful and commonly influence water chemistry. At Malaga Bend, a fissure in a contained saline aquifer seeps out at 0.4 ft^3/sec, containing approximately 250,000 ppm of TDS. This inflow doubles the TDS in the Pecos.

USES OF THE WATERSHED

Agriculture

The land use of the watershed in New Mexico is dominated by grazing (90%). Commercial timber (5.3%) and cultivated crops (2.3%) account for most of the rest. Cultivated crops are dominated by irrigated alfalfa, feed grains, and—to a lesser extent—cotton. Water availability is a primary factor limiting agriculture. Almost 85% of the water use in the basin is for irrigation (NMWQCC, 1976).

Cattle is the main use of the land in the Pecos valley. Some cotton is also grown. Small but intensive cantalope growing occurs in the Pecos valley as well as grain and sourgum that have increased in importance (NMWQCC, 1976).

Irrigation

Irrigation in Texas from the Pecos River between Red Bluff and Girvin began in 1877. Ten major irrigation projects affected this reach (by 1910), but technical difficulties, salinization, and undependable flow plagued the projects. Irrigation was expanded with the operation of Red Bluff Reservoir. Since 1940, groundwater of fair to marginal quality was used in Ward and Reeves counties (Figure 4.2) (Mendieta, 1974). Irrigation in the Texas part of the Pecos is mainly between Toyah Creek and Girvin. Most of this is from groundwater.

Salinity. Salt deposits in the Pecos valley between Malaga, New Mexico and Pecos, Texas produce brackish waters with conductivity averaging 14,000 μS cm^{-1} or greater at Girvin, Texas. Freshwater aquifer discharges supplement the flow downstream from Sheffield so that the stream is essentially a freshwater stream near its confluence with the Rio Grande.

Oil and Gas Fields

Oil and gas fields are extensive through most of the Pecos in Texas. Because much of the flow of the Pecos is naturally saline, detailed studies are needed to identify the sources of brine contributed by oil fields. In a short reach of about 3 mi in the Malaga Bend, about 420 tons of dissolved minerals, mainly NaCl, is added daily to the mineral load of the river. Water in the Pecos is very saline as it enters Texas. The source of the salt is in a concentrated brine that percolates upward from the aquifer (Mendieta, 1974).

Effects of Releases from Red Bluff Reservoir

Red Bluff Reservoir, which is the reservoir through which the Pecos River enters Texas, was started in 1937 and has a capacity of 310,000 acre-ft. It is used for irrigation. The Amistad Reservoir on the Rio Grande, which was completed in 1968, has a capacity of 5,325,000 acre-ft. This reservoir is used for irrigation, municipal water, industrial water, flood control, recreation, and hydroelectric power (Mendieta, 1974).

At Orla, periodic releases from Red Bluff Reservoir are indicated by the shifts in flow. Particularly between spring and summer there is a wide range in the amount of flow. This is caused partially by the need of water for irrigation. Due to the heavy

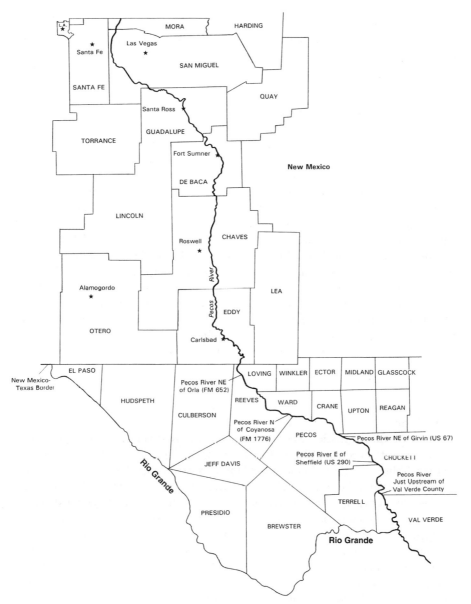

FIGURE 4.2.   Map of Pecos River basin counties. (Created by Susan Durdu; drawn by Su-Ing Yong.)

withdrawals of water for irrigation the streambed is relatively scoured from bank to bank (TNRCC, 1996a).

The only instream cover is provided by the dense riparian vegetation of salt cedar. It grows out into the water surface. The high flows of the river cause the stream banks to be somewhat steep. This high flow is caused by releases from Red Bluff Reservoir, approximately 14 mi upstream. Usually, the streambed is very firm, which seems to be the result of scouring; the only places where depositional material accumulate are at an occasional sandbar, where rocky aggregations might be found.

Pollution

Nonsource runoff from cropland, ranges, and pastures is a major mode of pollution entry for the Pecos. Feedlots contribute to nitrate, ammonium, and organic carbon in groundwater. Artesia and Carlsbad are the major point sources of municipal waste to the Pecos in New Mexico (NMWQCC, 1976).

An experimental Salinity Alleviation Project was started in the Malaga Bend area in August 1963. The Pecos receives better water from the Delaware, but the influx is small. The Delaware water is very hard and dominated by calcium sulfate. Evaporation, irrigation returns, and the inflow of highly mineralized groundwater and oilfield brines usually cause a progressive increase in total dissolved solids (TDS) and chloride in the Pecos from Red Bluff to Girvin. Farther downstream, the inflow of groundwater usually results in a reduction of TDS. For instance, the total dissolved solids from 1961 to 1968 averaged 7770 mg L^{-1} below Red Bluff, 14,200 mg L^{-1} at Girvin, and 1600 mg L^{-1} at Shumla. (There was a great reduction between Girvin and Shumla.) The discharge weighted concentrations of chloride were 3080; 5820; and 572 mg L^{-1}, respectively. The discharge weighted mean concentrations for sulfate were 3370 mg L^{-1} at Girvin and 324 mg L^{-1} at Shumla (Mendieta, 1974).

CHEMICAL CHARACTERISTICS

New Mexico

Trace Metals.   Studies were made of the Pecos River approximately 400 m above the confluence with Willow Creek (upper Pecos) in 1980–1981 by Jacobi and Smolka (Table 4.1). From October 1980 through November 1981 the water temperature varied from 1.0 to 13.8°C. The conductivity at 25°C varied from 105 to 149 µS. The dissolved oxygen varied from 8.0 to 10.6 mg L^{-1} throughout the year and the percent oxygen saturation varied from 94 to 105%. pH varied from 7.6 to 8.00. The total nitrogen (ammonia plus nitrates) varied from 0.005 to 0.250 mg L^{-1} and total Kjeldahl nitrogen varied from 0.090 to 0.670 mg L^{-1}. Total nitrogen in terms of nitrites and nitrates varied from 0.01 to 0.06 mg L^{-1}. Total phosphorus varied from 0.002 to 0.260 mg L^{-1} (Jacobi and Smolka, 1982). Trace metals were determined by Jacobi and Smolka (1982) and found to be very low (Table 4.1).

The results obtained by Krauskopf (1979) from roughly the same locality were as follows. Mercury was 90 to 110 ppb, depending on methods of analysis. Lead was 300 ppb on particles less than 75 µg in size and 255 ppb for those 75 µm or more in size. Copper was 310 ppm on particles less than 75 µm. Copper on particles greater than 75 µm was 250 ppm. Zinc on particles less than 75 µm was 3100 ppm and on particles greater than 75 µm was 2900 ppm. Chromium was less than 62 ppm on particles less than 75 µm and on particles greater than 75 µm it was 130 ppm (McLemore et al., 1993).

TABLE 4.1. *Water Quality Data, October 1980–November 1981, Pecos River 400 m above Confluence with Willow Creek*

Date	Time of Day	Depth (ft.)	Water Temp. °C	Conductivity at 25°C μS	Dissolved Oxygen (mg L⁻¹)	Dissolved Oxygen Percent Saturation	pH (standard units)	Total NH₃ + NH₄⁻ N (mg L⁻¹)	Total Kjeldahl N (mg L⁻¹)	Total NO₂ and NO₃⁻ N (mg L⁻¹)	Total Phosphate (mg L⁻¹ P)	HCO₃ Ions (mg L⁻¹)
10/30/80	13:00	0	4.5	162	10.0	103.0	7.90	0.011	0.270	0.03	0.002K	—
12/4/80	11:00	0	1.0	149	—	—	7.80	0.070	0.540	0.06	0.011	96
1/9/81	12:00	30	1.0	147	—	—	7.60	0.020	0.110	0.04	0.005K	95
2/24/81	12:00	15	1.8	139	10.6	101.0	7.80	0.070	0.190	0.04	0.002	90
4/8/81	13:00	0	8.9	149	9.1	105.0	7.80	0.053	0.590	0.03	0.002K	87
5/20/81	13:00	0	8.5	110	8.7	100.0	7.70	0.100	0.140	0.01	0.010	64
6/26/81	10:00	50	13.8	143	8.0	104.0	8.00	0.250	0.670	0.01K	0.260	81
8/13/81	11:00	45	10.7	105	8.1	98.0	7.60	0.005	0.460	0.01	0.020	53
10/7/81	10:00	45	7.5	115	8.5	94.0	7.70	0.020	0.130	0.01K	0.005	70
11/24/81	11:00	30	3.0	144	9.7	97.0	7.80	0.041	0.090	0.04	0.030	87
Station: Number			10	10	8	8	10	10	10	10	10	9
Maximum			13.800	162.000	10.600	105.000	8.000	0.250	0.670	0.060	0.260	96.100
Minimum			1.000	105.000	8.000	94.000	7.600	0.005	0.090	0.010	0.002	52.700
Mean			6.070	136.300	9.087	100.250	7.770	0.064	0.319	0.027	0.035	80.278
Std. dev.			4.456	19.247	0.937	3.770	0.125	0.072	0.223	0.017	0.080	14.923
Std. error			1.409	6.087	0.331	1.333	0.040	0.023	0.071	0.005	0.005	4.974

(*continued*)

TABLE 4.1. (Continued)

Date	Time of Day	Depth (ft)	Total Vanadium (μG L^{-1})	Zinc Diss. (μG L^{-1})	Zinc Total (μG L^{-1})	Selenium Diss. (μG L^{-1})	Selenium Total (μG L^{-1})	Total Natural Uranium (μG L^{-1})	Mercury Diss. (μG L^{-1})	Mercury Total (μG L^{-1})	Barium Diss. (μG L^{-1})	Barium Total (μG L^{-1})
10/30/80	13:00	0	—	—	—	5K	—	—	0.5K	—	100K	—
12/4/80	11:00	0	—	250K	—	5K	—	—	0.5K	—	100K	—
1/9/81	12:00	30	10K	—	250K	—	5K	12.0	—	0.5K	—	270
2/24/81	12:00	15	—	—	100K	—	5K	—	—	0.5K	—	100K
4/8/81	13:00	0	—	—	100K	—	5K	—	—	0.5K	—	100K
5/20/81	13:00	0	—	—	30K	—	5K	—	—	0.5K	—	100K
6/26/81	10:00	50	10K	—	100K	—	5K	5.0K	—	0.5	—	100K
8/13/81	11:00	45	—	—	100K	—	—	—	—	—	—	100K
10/7/81	10:00	45	—	—	100K	—	5K	—	—	0.5K	—	100K
11/24/81	11:00	30	10K	—	100K	—	5K	—	—	0.5K	—	100
Station: Number			3	1	8	2	7	2	2	8	2	8
Maximum			10.000	250.000	250.000	5.000	5.000	12.000	0.500	0.500	100.000	270.000
Minimum			10.000	250.000	30.000	5.000	5.000	5.000	0.500	0.500	100.000	100.000
Mean			10.000	250.000	110.000	5.000	5.000	8.500	0.500	0.500	100.000	121.250
Std. dev.			0.000	—	61.644	0.000	0.000	4.950	0.000	0.000	0.000	60.104
Std. error			0.000	—	21.794	0.000	0.000	3.500	0.000	0.000	0.000	21.250

TABLE 4.1. *(Continued)*

Date	Time of Day	Depth (ft)	Berylium Diss. (μG L⁻¹)	Berylium Total (μG L⁻¹)	Cadmium Diss. (μG L⁻¹)	Cadmium Total (μG L⁻¹)	Chromium Diss. (μG L⁻¹)	Chromium Total (μG L⁻¹)	Copper Diss. (μG L⁻¹)	Copper Total (μG L⁻¹)	Iron Diss. (μG L⁻¹)	Iron Total (μG L⁻¹)
10/30/80	13:00	0	—	—	1K	—	5K	—	—	—	—	—
12/4/80	11:00	0	10.00K	—	1K	—	5K	—	10K	—	—	250K
1/9/81	12:00	30	—	10.00K	—	1K	—	5K	—	10K	250K	—
2/24/81	12:00	15	—	10.00K	—	1K	—	5K	—	10K	250K	—
4/8/81	13:00	0	—	10.00K	—	1K	—	5K	—	10K	250K	—
5/20/81	13:00	0	—	10.00K	—	1K	—	5K	—	10K	30	—
6/26/81	10:00	50	—	10.00K	—	1	—	5K	—	50K	38	—
8/13/81	11:00	45	—	10.00K	—	1K	—	5K	—	50K	460	—
10/7/81	10:00	45	—	1.00K	—	1	—	5K	—	100K	100K	—
11/24/81	11:00	30	—	1.00K	—	1K	—	5K	—	100K	100K	—
Station: Number			1	8	2	8	2	8	1	8	8	1
Maximum			10.000	10.000	1.000	1.000	5.000	5.000	10.000	100.000	460.000	250.000
Minimum			10.000	1.000	1.000	1.000	5.000	5.000	10.000	10.000	30.000	250.000
Mean			10.000	7.750	1.000	1.000	5.000	5.000	10.000	42.500	184.750	250.000
Stand. dev.			—	4.166	0.000	0.000	0.000	0.000	—	39.551	145.594	—
Stand. error			—	1.473	0.000	0.000	0.000	0.000	—	13.983	51.4753	—

(continued)

TABLE 4.1. (*Continued*)

Date	Time of Day	Depth (ft)	Lead Diss. (μG L⁻¹)	Lead Total (μG L⁻¹)	Manganese Diss. (μG L⁻¹)	Manganese Total (μG L⁻¹)	Molybdenum Diss. (μG L⁻¹)	Molybdenum Total (μG L⁻¹)	Silver Diss. (μG L⁻¹)	Silver Total (μG L⁻¹)
10/30/80	13:00	0	5K	—	—	—	5K	—	—	—
12/4/80	11:00	0	5K	—	—	—	—	—	1.0K	—
1/9/81	12:00	30	—	5K	100.0K	—	—	5K	100.0K	1.0K
2/24/81	12:00	15	—	5K	50.0K	—	—	5K	—	—
4/8/81	13:00	0	—	5K	50.0K	—	—	5K	—	1.0K
5/20/81	13:00	0	—	5K	50.0K	—	—	5K	—	1.0K
6/26/81	10:00	50	—	5K	50.0K	—	—	5K	—	1.0K
8/13/81	11:00	45	—	5K	50.0K	—	—	5K	—	1.0K
10/7/81	10:00	45	—	5K	50.0K	—	—	10K	—	1.0K
11/24/81	11:00	30	—	5K	50.0K	—	—	10K	—	1.0K
Station: Number			2	8	8	—	1	8	2	7
Maximum			5.000	5.000	100.000	—	5.000	10.000	100.000	1.000
Minimum			5.000	5.000	50.000	—	5.000	5.000	1.000	1.000
Mean			5.000	5.000	56.250	—	5.000	6.250	50.500	1.000
Std. dev.			0.000	0.000	17.678	—	—	2.315	70.004	0.000
Std. error			0.000	0.000	6.250	—	—	0.818	49.500	0.000

Source: Modified from Jacobi and Smolka (1982).

A study was made by Forest Hawman (1995) of the heavy metal concentrations in sediments and fish in the Pecos River in New Mexico. These studies were also made in the Santa Rosa Reservoir. Samples collected from the Pecos River in sediments at Santa Rosa, New Mexico below the reservoir indicated concentrations of the following substances: copper 22 ppm, lead 51 ppm, cadmium 2 ppm, chromium 34 ppm, zinc 49 ppm, nickel 44 ppm, and mercury 0.02 ppm. These studies were carried out in 1995. Higher concentrations were found in most cases in the reservoir (Hawman, 1995).

In Santa Rosa Reservoir mercury concentrations were analyzed in the bottom sediments, plankton, small fish, and large predator fish. Hawman found the highest average concentrations of total metals in Santa Rosa Reservoir sediments at Los Esteros (260 mg L^{-1}) (Figure 4.3), indicating that the suspended sediments that settled in this site were composed of the optimum-size particles for metal absorption. They found that sediments high in organic carbon tended to settle in this site. The East Bay site was located in a shallow area, where the flow of water was slow and metal concentrations would accumulate in the base soils. Reuther (1994) reported the following concerning mercury found in Santa Rosa Reservoir: Bottom sediments were about 20 ppb. Plankton probably had about 200 ppb dry weight. In small fish the concentration was 400 to 1000 ppb dry weight and in large predator fish mercury was as high as 1500 to 2700 ppb dry weight. Similar results were found by Reuther (1994) in the Rio Mutum Paraná (Brazil) (Hawman, 1995).

Although the water and sediments in Santa Rosa Reservoir did not contain much mercury, the fish showed the presence of mercury in their tissues and there seemed

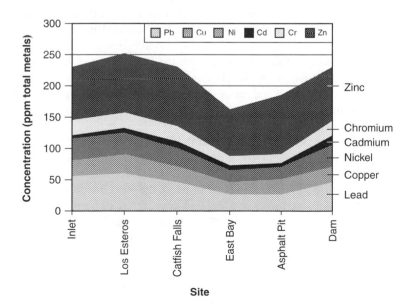

FIGURE 4.3.   Average total metal concentrations for all sites within the Santa Rosa reservoir (Hawman, 1995).

to be a strong correlation between mercury concentrations and body weight. This is similar to a report on mercury of the U.S. Bureau of Reclamation in 1992 (McLemore et al., 1993). This report also showed elevated levels of mercury in fish in Santa Rosa Reservoir. In 58% of the fish obtained for this study (11 of the 19 fish studied) mercury concentrations in their muscle tissue were in excess of 1 ppm dry weight, which is the limit established by the Environmental Protection Agency. This indicates a possible mercury contamination source exists, although data obtained for this study did not give any indication as to the origin of the contamination. Native trout collected in the upper Pecos River upstream from the mine area did not show any sign of mercury contamination (the limit of determination is approximately 0.2 ppm dry weight). This indicates either that there is no source of mercury in the upper Pecos or that mercury is not bioavailable in this environment (Hawman, 1995). However, this does not eliminate the fact that tributaries may bring heavy metals into the Pecos River (Hawman, 1995).

Analysis of sediments above and below the reservoir indicated that the reservoir is not having an adverse effect on the river downstream by releasing sediments that have been contaminated with high concentrations of heavy metals. It appears that the mercury and other metals were the result of releases from the mines upstream from the reservoir (Hawman, 1995).

Sediment and water samples obtained from the Santa Rosa Reservoir and the Pecos River at Tecolote to the gauging station show little sign of contamination (Hawman, 1995). Total concentrations in sediments of the Pecos River near Santa Rosa Reservoir were consistently lower than those found in the upper Pecos River above the Terrerro mine (Looper, 1995), indicating that these metals are not reaching the reservoir in any substantial quantities. The metal concentrations in the sediments are comparable to those found in other noncontaminated rivers in New Mexico, such as the Red River (located on the border of Texas and Oklahoma and in Arkansas and Louisiana) (Lynch et al., 1988), the Rio Salado, and the Rio Puerco (Brandvold et al., 1981; Hawman, 1995).

The results of studies of samples taken from the Pecos River at the Pecos Wilderness area (the southern tip of the Sangre de Cristo Mountains) southward to north of Carlsbad indicated that lead and zinc concentrations in Pecos River water samples were above average dissolved concentrations found in the stream water, whereas copper was lower (Hawman, 1995).

McLemore et al. (1993) also concluded that lead and zinc concentrations in the Pecos water samples were above average dissolved concentrations found in stream water, whereas copper was lower. When the Pecos concentrations were compared with those found in another river in New Mexico, the Rio Grande, copper and mercury concentrations were lower and the lead concentrations were about the same. Zinc concentrations appeared slightly elevated in the upper Pecos. Comparing the Pecos concentrations with New Mexico Water Quality Control Commission District Stream Standards, the area just below the confluence with Willow Creek was one of two sites with concentrations above the standards for lead and zinc only. Conductivity increased downstream in the Pecos River, possibly a natural result from Permian evaporates and limestone, contamination from agricultural areas, as well as evaporation (McLemore et al., 1993).

Mercury and zinc concentrations in sediments were typically below crustal (concentration in earth surface) abundance or below averages for common lithologies found in the Pecos drainage (Krauskopf, 1979), with the exception of Jacks Creek just below the village of Pecos. Copper and chromium concentrations in sediments were below crustal abundances except for samples from the Pecos mine water drainage. Mercury, lead, copper, and zinc concentrations in sediments were elevated in samples from above the Pecos mine waste dumps, suggesting that outcroppings, zones of mineralization, and perhaps the outcropping rocks are sources of contamination as well as the waste dumps themselves (McLemore et al., 1993).

Lead concentrations in the Pecos River sediments were above crustal abundance. Lead concentrations in the river sediments elsewhere in New Mexico are also typically above crustal abundance [Brandvold and Brandvold, 1990; U.S. Geological Survey (USGS) et al., 1980]. Some studies indicate that lead and mercury contamination in the atmosphere are increased with time as a result of human-induced pollution. A major source of lead into the atmosphere has been automobile exhaust from lead gasoline. Precipitation washes some of these pollutants from the atmosphere into the water supply and soils. It is also possible that rocks and sediments throughout New Mexico may be enriched in lead relative to crustal abundance (McLemore et al., 1993).

Concentrations of metals in the Pecos River sediments were typically higher in lake sediments than in stream sediments, with the above-mentioned exceptions. The concentrations of metals in the bottom sediment fraction (75–2 μg) are typically lower than the fraction below 75 μg. Clay-sized fractions tend to have higher concentrations of metals than sand and silt. Apparently, even though mercury, lead, copper, and zinc concentrations were elevated in some sediments, the concentrations in sediments have little, if any effect on metal concentrations in the water. However, these metals in sediments may enter the food chain and eventually concentrate in fish. Bioaccumulation of metals is a very complex issue and is not the focus of the report. White and brown precipitates or froth from the seep below the waste dumps were high in zinc and other metals. During high-runoff events, it is possible that the precipitate enters in the Pecos River and subsequently may enter the fish hatchery at Lisbon Springs. Fish kills have been reported in part as a result of contamination by precipitates during high-runoff events. However, a U.S. Fish and Wildlife Services report suggests that the fish kills at Lisbon Springs State Fish Hatchery are due to a combination of water quality factors, including temperature, periodic presence of the toxic contamination of metals, and the possible introduction of disease (*Albuquerque Journal of Fish*, vol. 16, 1993). Furthermore, fish along the Pecos River between the mine dumps and the factory (a distance of 19 km) appear healthy and show no effects from potential contaminations (McLemore et al., 1993).

Texas

Orla to Shumla.   At Orla, Sheffield, and Shumla the alkalinity ranged between 100 and 200 mg L^{-1}. At Girvin it ranged from 50 to 200 mg L^{-1}. pH ranged from 7.8 to 8.5 at Orla, Sheffield, and Shumla, but the water was hyperalkaline at Girvin (associated with the low bicarbonate alkalinity). Alkalinity at Girvin was low during primary production. The increase in pH (>10) resulted in carbonate precipitation.

Specific conductance ranged approximately between 10,000 and 20,000 µS cm^{-1} at Orla. It was somewhat higher at Girvin and ranged between 5000 and 15,000 at Sheffield and was less than 3000 at Shumla. Sulfates were typically in excess of 1000 mg L^{-1} at the three upstream stations but less than 500 mg L^{-1} at Shumla (Davis, 1980).

Total organic carbon ranged from about 6 to 12 mg L^{-1} at Orla, Girvin, and Sheffield but was less than 1 to 7 mg L^{-1} at Shumla. Water column chlorophyll *a* usually ranged between 0.005 and 0.010 mg L^{-1}. Total organic carbon amounts at Orla and Girvin were elevated because of allochthonous inputs (probably from salt cedar) in the upper station (Orla) and autochthonous inputs at Girvin. Seasonally, TOC is higher in the winter, when allochthonous salt cedar inputs are high.

Nitrate nitrogen was high (2.7 mg L^{-1}), but it averaged approximately 0.6 mg L^{-1} at Orla. At Girvin, nitrogen as NO_3 was less than 0.5 mg L^{-1} and at Sheffield it ranged from 0.3 to 0.8 mg L^{-1}. Increased groundwater inputs elevated nitrogen as NO_3 at Shumla to a range between 1 and 2.6 mg L^{-1}. Ammonia nitrogen was high at Orla (0 to 1.5 mg L^{-1}) and at Girvin (0.1 to 0.9 mg L^{-1}) with seasonal peaks in the winter which were associated with autotrophic senescence and allochthonous inputs. Autotrophic depletion appears to have reduced levels by the time the river reached Sheffield (<0.2 mg L^{-1}) and the prevalence of groundwater inflow limited ammonia nitrogen at Shumla to less than 0.2 mg L^{-1} (Davis, 1980).

Total phosphorus at Orla and Girvin exhibited winter peaks associated with autotrophic senescence and allochthonous inputs. The downstream stations (Sheffield and Shumla) yielded relatively low concentrations (less than 0.04 mg L^{-1}) presumably because of the predominance of spring flow. Total organic carbon and chlorophyll *a* were associated with phosphorus, which suggests that autochthonous primary production was phosphorus limited (Davis, 1980).

Dissolved oxygen was typically greater than 5 mg L^{-1}. There was considerable fluctuation at Orla and Girvin. Dense aquatic vegetation reduced dissolved oxygen to 3.7 mg L^{-1} at Girvin (approximately 42%) in April because of high respiration demand (taken at 10:00 under cloudy conditions). Mean stream temperature increased downstream, but temperatures were more variable upstream and the maximums were higher upstream. The main difference is that the minimums were higher at Shumla.

Turbidity was relatively high at all stations but highest at Orla and lowest at Shumla. Salinity reduced turbidity at Girvin by precipitating suspended solids. Discharge remained slow (Table 4.2), less than 75 m^3 s^{-1} at the three upper stations but ranged from 200 to 450 m^3 s^{-1} at Shumla (Davis, 1980).

September 1994. From an analysis done in September 1994, flow (ft^3/s) at Orla was 20; at Coyanosa 12; Girvin 25; Sheffield 34; and at Shell Slab 77. The depth of the water is 0.3 m at Orla, 0.3 m at Coyanosa, 0.3 m at Girvin, 0.3 m at Sheffield, and 0.3 m at Shell Slab. The water temperature was 20°C at Orla, 25°C at Coyanosa, 23.3°C at Girvin, 25.2°C at Sheffield, and 24.8°C at Shell Slab, thus showing a slight increase in temperature as one proceeds downstream. The pH was 7.9 at Orla, 7.3 at Coyanosa, 7.9 at Girvin, 7.8 at Sheffield, and 7.6 at Shell Slab. Dissolved oxygen (mg L^{-1}) was 7.3 ppm at Orla, 6.6 ppm at Coyanosa, 9.2 ppm at Girvin, 7.2 ppm at Sheffield, and 8.4 ppm at Shell Slab. Conductivity was 12,600 µS cm^{-1} at Orla,

TABLE 4.2. *Monthly Mean Discharge in the Middle and Lower Pecos River, New Mexico and Texas, Water Year 1964*

	Ft. Sumner, NM	Artesia, NM	Malaga, NM	Orla, TX	Girvin, TX
Oct.	44	21	24	19	11
Nov.	27	46	40	11	20
Dec.	18	54	51	12	24
Jan.	14	44	42	8	24
Feb.	14	46	34	42	24
Mar.	28	31	21	49	51
Apr.	50?	299	18	10	17
May	30	18	15	86	9
June	25?	58	23	113	8
July	35?	164	10	13	4
Aug.	23?	<1	8	25	6
Sept.	32	<1	18	29	24
Water year mean		65	25	35	18
Maximum	121	1000	266	278	161
Minimum	12	0	7	1	2

Source: Modified from USGS (1970).

15,700 µS cm^{-1} at Coyanosa, 16,400 µS cm^{-1} at Girvin, 14,100 µS cm^{-1} at Sheffield, and 7350 µS cm^{-1} at Shell Slab. This indicates the shift in water conductivity between Sheffield and Shell Slab.

Turbidity as measured by depth in meters with a Secchi disk was 0.87 m at Orla, 0.59 m at Coyanosa, >1.0 m at Girvin, 0.94 m at Sheffield, and 0.87 m at Shell Slab. Thus it is evident that the clearness of the water decreases as one goes downstream. Values at Girvin, Sheffield, and Shell Slab were higher than at Coyanosa and not much different from that at Orla. Values were the same at Shell Slab and Orla.

Pecos River Water near Langtry. Chemical analysis of the water was made for the calendar year January–December 1999 in the Pecos River in the vicinity of Langtry (Gibbons, 2001) (Table 4.3). Discharge varied from 84 ft^3/s in September to 157 ft^3/s in January. Residual solids at 180°C varied from 1700 mg L^{-1} in September to 2570 mg L^{-1} in April, and dissolved solids varied from 1700 mg L^{-1} in September to 2510 mg L^{-1} in April. Turbidity varied from 0.3 NTU in January to 1.8 NTU in March. Dissolved oxygen varied from 7.1 mg L^{-1} in June and August to 10.6 mg L^{-1} in January. Percent dissolved oxygen varied from 100% in January to 113% in April. pH varied from 7.9 in May to 8.2 in March, April, June, and September. Specific conductivity ranged from 2880 µS cm in June to 4060 µS cm in April. Temperature ranged from 10.0°C in January to 30.7°C in June and August. Hardness in terms of calcium carbonate per liter varied from 440 mg L^{-1} in September to 700 mg L^{-1} in April. Hardness as calcium carbonate varied from 540 mg L^{-1} in September to 820 mg L^{-1} in April.

TABLE 4.3. *Water Quality Data, Calendar Year January–December 1999 Pecos River near Langtry, Texas*

Date	Time	Instantaneous Discharge ft³/s	Solids Dissolved Residue at 180°C (mg L⁻¹)	Solids Sum of Dissolved Constituents, (mg L⁻¹)	Turbidity (NTU)	Dissolved Oxygen (mg L⁻¹)	Dissolved Oxygen Percent saturation	pH of Water, Whole Field (standard units)	Specific Conductance (µS cm⁻¹)	Water Temperature (°C)	Hardness of Non-carbonate Dissolved Fluid as CaCO₃ (mg L⁻¹)
Jan. 12	14:30	157	2390	2310	0.3	10.6	100	8.0	3900	10.0	650
Mar. 23	14:40	151	2130	2050	1.8	9.0	110	8.2	3550	21.2	570
Apr. 20	15:30	112	2570	2510	1.2	8.9	113	8.2	4060	24.0	700
May 18	14:50	139	2260	2170	1.5	7.6	105	7.9	3670	27.8	610
June 29	14:10	131	1790	1720	1.5	7.1	101	8.2	2880	30.7	450
Aug. 2	15:00	92	1880	1810	1.0	7.1	102	8.1	3060	30.7	480
Sept. 8	15:20	84	1700	1700	0.4	7.3	104	8.2	2890	29.6	440

Date	Time	Total Hardness (mg L⁻¹ as CaCO₃)	Dissolved Calcium (mg L⁻¹ as Ca)	Dissolved Magnesium (mg L⁻¹ as Mg)	Dissolved Potassium (mg L⁻¹ as K)	Sodium Adsorption Ratio	Dissolved Sodium (mg L⁻¹ as Na)	Alkalinity Water Total lt. Field (mg L⁻¹ as CaCO₃)	Dissolved Bicarbonate in Water, lt. Field (mg L⁻¹ as HCO₃)	Dissolved Carbonate in Water, lt. Field (mg L⁻¹ as CO₃)	Dissolved Chloride, (mg L⁻¹ as Cl)	Dissolved Fluoride, (mg L⁻¹ as F)
Jan. 12	14:30	810	181	86.0	8.6	8	520	160	195	0	873	0.8
Mar. 23	14:40	700	153	75.6	7.3	8	460	125	152	0	791	0.8
Apr. 20	15:30	820	173	91.9	8.5	9	568	111	136	0	1020	0.9
May 18	14:50	710	149	80.5	8.3	8	514	100	122	0	848	0.9
June 29	14:10	550	117	62.9	7.0	7	392	104	127	0	673	0.8
Aug. 2	15:00	580	120	66.7	7.7	8	424	100	122	0	701	0.8
Sept. 8	15:20	540	110	63.3	6.4	7	393	98	119	0	668	0.9

TABLE 4.3. (Continued)

Date	Time	Dissolved Silica (mg L^{-1} as SiO$_2$)	Dissolved Sulfate (mg L^{-1} as SO$_4$)	Dissolved Ammonia (mg L^{-1} as N)	Dissolved Ammonia + Organic (mg L^{-1} as N)	Total Ammonia + Organic (mg L^{-1} as N)	Nitrogen Dissolved Nitrate (mg L^{-1} as N)	Dissolved NO$_2$ + NO$_3$ (mg L^{-1} as N)	Dissolved Nitrite (mg L^{-1} as N)	Dissolved Organic (mg L^{-1} as N)	Total Organic (mg L^{-1} as N)	Total (mg L^{-1} as N)
Jan. 12	14:30	13.8	528	<0.020	0.10	0.13	—	0.797	<0.010	—	—	0.93
Mar. 23	14:40	10.8	473	<0.020	0.18	0.23	—	0.435	<0.010	—	—	0.66
Apr. 20	15:30	11.3	563	0.030	0.18	0.23	—	0.196	<0.010	0.15	0.20	0.43
May 18	14:50	12.3	493	0.049	0.10	0.30	—	0.067	<0.010	—	0.25	0.37
June 29	14:10	9.8	387	0.069	0.28	0.31	0.099	0.111	0.012	0.21	0.25	0.43
Aug. 2	15:00	12.9	415	0.039	0.23	0.30	—	<0.050	<0.010	0.19	0.26	—
Sept. 8	15:20	13.2	380	0.027	0.20	0.12	—	<0.050	<0.010	0.17	0.09	—

Date	Time	Dissolved Ortho-phosphate (mg L^{-1} as PO$_4$)	Phosphorus Dissolved (mg L^{-1} as P)	Dissolved Ortho (mg L^{-1} as P)	Total (mg L^{-1} as P)	Carbon Dissolved Organic (mg L^{-1} as C)	Total Organic Particulate (mg L^{-1} as C)	Dissolved Aluminum (µg L^{-1} as Al)	Dissolved Antimony (µg L^{-1} as Sb)	Dissolved Arsenic (µg L^{-1} as As)	Dissolved Barium (µg L^{-1} as Ba)	Dissolved Beryllium (µg L^{-1} as Be)
Jan. 12	14:30	0.006	<0.050	0.002	<0.050	1.2	<0.2	—	—	1.2	—	—
Mar. 23	14:40	0.003	<0.004	0.001	0.010	2.5	<0.2	2	<2	<1.0	76	<2
Apr. 20	15:30	0.006	<0.004	0.002	<0.004	2.0	<0.2	—	—	1.6	—	—
May 18	14:50	0.009	<0.004	0.003	<0.004	2.8	<0.2	8	<2	<1.0	77	<2
June 29	14:10	0.006	<0.004	0.002	<0.004	4.7	0.3	—	—	1.2	—	—
Aug. 2	15:00	0.003	<0.004	0.001	0.006	1.3	0.2	—	—	1.5	—	—
Sept. 8	15:20	—	<0.004	<0.001	<0.004	2.3	0.2	<2	<2	1.3	70	<2

(continued)

TABLE 4.3. (Continued)

Date	Time	Dissolved Boron (µg L⁻¹ as B)	Dissolved Cadmium (µg L⁻¹ as Cd)	Dissolved Chromium (µg L⁻¹ as Cr)	Dissolved Cobalt (µg L⁻¹ as Co)	Dissolved Copper (µg L⁻¹ as Cu)	Dissolved Iron (µg L⁻¹ as Fe)	Dissolved Lead (µg L⁻¹ as Pb)	Dissolved Lithium (µg L⁻¹ as Li)	Dissolved Manganese (µg L⁻¹ as Mn)	Dissolved Molybdenum (µg L⁻¹ as Mo)
Jan. 12	14:30	150	—	—	—	—	<20	—	76.4	—	—
Mar. 23	14:40	216	<2.0	5.2	<2	<2	<30	<2	64.8	2	7
Apr. 20	15:30	266	—	—	—	—	<30	—	66.5	6	7
May 18	14:50	970	<2.0	<1.0	<2	3	<30	<2	83.0	—	—
June 29	14:10	221	—	—	—	—	<30	—	60.6	—	—
Aug. 2	15:00	235	—	—	—	—	<30	—	68.8	—	—
Sept. 8	15:20	229	<2.0	<1.0	<2	<2	<30	<2	66.5	2	6

Suspended Sediment

Date	Time	Dissolved Nickel (µg L⁻¹ as Ni)	Dissolved Selenium (µg L⁻¹ as Se)	Dissolved Silver (µg L⁻¹ as Ag)	Dissolved Strontium (µg L⁻¹ as Sr)	Dissolved Vanadium (µg L⁻¹ as V)	Dissolved Zinc (µg L⁻¹ as Zn)	Natural Dissolved Uranium (µg L⁻¹ as U)	% Finer than 0.062 mm Sieve Diam.	(mg L⁻¹)	Discharge (tons/day)
Jan. 12	14:30	—	1.4	—	3330	21	—	—	83	<1	—
Mar. 23	14:40	4	1.2	<2	2810	19	2	4	93	3	1.2
Apr. 20	15:30	—	<1.0	—	3230	26	—	3	75	2	0.60
May 18	14:50	5	<1.0	<2	2980	21	5	3	62	2	0.75
June 29	14:10	—	<1.0	—	2330	15	—	—	93	2	0.71
Aug. 2	15:00	—	<1.0	—	2440	18	—	—	76	3	0.75
Sept. 8	15:20	3	<1.0	<2	2340	16	<2	2	80	1	0.23

Source: Modified from Gibbons (2001).

Calcium varied from 110 mg L^{-1} in September to 181 mg L^{-1} in January. Magnesium varied from 62.9 mg L^{-1} in June to 91.9 mg L^{-1} in April. Potassium varied from 6.4 mg L^{-1} in September to 8.6 mg L^{-1} in January. Sodium (dissolved) varied from 392 mg L^{-1} in June to 568 mg L^{-1} in April. Alkalinity varied from 98 mg L^{-1} in September to 160 mg L^{-1} in January. Bicarbonates varied from 119 mg L^{-1} in September to 195 mg L^{-1} in January. Chlorides varied from 668 mg L^{-1} in September to 1020 mg L^{-1} in April. Silica varied from 9.8 mg L^{-1} in June to 13.8 mg L^{-1} in January. Dissolved sulfates varied from 380 mg L^{-1} in September to 563 mg L^{-1} in April.

Nitrogen as ammonia varied from <.020 mg L^{-1} in January and March to 0.069 mg L^{-1} in June. Organic nitrogen varied from 0.10 mg L^{-1} in January to 10 mg L^{-1} in May. Total nitrogen as ammonia plus organics varied from 0.12 mg L^{-1} in September to 0.31 mg L^{-1} in June. Nitrogen as nitrates was recorded only once, in June, and it was 0.099 mg L^{-1}. Nitrogen as $NO_2 + NO_3$ varied from <0.05 mg L^{-1} in August and September to 0.797 mg L^{-1} in January. Nitrogen as dissolved nitrites varied from <0.010 mg L^{-1} in January, March, April, May, and August and September to 0.012 mg L^{-1} in June. Organic nitrogen varied from 0.15 mg L^{-1} in April to 0.21 mg L^{-1} in June. Total organic nitrogen varied from 0.09 mg L^{-1} in September to 0.26 mg L^{-1} in August. Phosphate phosphorus as PO_4 varied from 0.003 mg L^{-1} in March and August to 0.009 mg L^{-1} in May. Dissolved phosphorus as P was always <0.004 mg L^{-1} except in January, when it was <0.050 mg L^{-1}. Orthophosphate dissolved was <0.001 mg L^{-1} in September to 0.003 in May. Total phosphorus as P varied from <0.004 mg L^{-1} in April, May, June, and September to 0.010 mg L^{-1} in March. Dissolved carbon as dissolved organic carbon varied from 1.2 mg L^{-1} in January to 4.7 mg L^{-1} in June. Particulate carbon was usually <0.2 mg L^{-1} but was as high as 0.3 mg L^{-1} in June.

Dissolved aluminum was only found three times, and it varied from <2 µg L^{-1} in September to 8 µg L^{-1} in May. Antimony (dissolved) was always <2 µg L^{-1}. Arsenic (dissolved) varied from <1.0 µg L^{-1} in March and May to 1.6 µg L^{-1} in April. Dissolved barium varied from 70 µg L^{-1} in September to 77 µg L^{-1} in May. Dissolved berylium was always <2 µg L^{-1}. In the calendar year 1999, dissolved boron varied from 150 µg L^{-1} in January to 970 µg L^{-1} in May. Dissolved cadmium was always <2 µg L^{-1}. Chromium (dissolved) varied from <1.0 µg L^{-1} in May and September to 5.2 µg L^{-1} in March. Cobalt was always <2 µg L^{-1}. Dissolved copper varied from <2 µg L^{-1} in March and September to 3 µg L^{-1} in May. Dissolved iron varied from <20 µg L^{-1} in January to <30 µg L^{-1} from March through September. Dissolved lead was recorded only three times and it was <2 µg L^{-1} in March, May, and September. Dissolved lithium varied from 60.6 µg L^{-1} in June to 83.0 µg L^{-1} in May. Dissolved manganese was recorded only three times and varied from 2 µg L^{-1} in March and September to 6 µg L^{-1} in May. Dissolved molybdenum varied from 6 µg L^{-1} in September to 7 µg L^{-1} in March and May. Dissolved nickel was recorded only three times and varied from 3 µg L^{-1} in September to 5 µg L^{-1} in May. Dissolved selenium varied from <1.0 µg L^{-1} in April through September to 1.4 µg L^{-1} in January. Dissolved silver was always <2 µg L^{-1}. Dissolved strontium varied from 2330 µg L^{-1} in June to 3330 µg L^{-1} in January. Dissolved vanadium varied from 15 µg L^{-1} in June to 26 µg L^{-1} in April. Dissolved zinc was recorded three times and it was <2 µg L^{-1} in September to 5 µg L^{-1} in May. Uranium (dissolved) varied from 2 µg L^{-1} in September to 4 µg L^{-1} in March. Suspended sediments varied from <1 mg L^{-1} in January to 3 mg L^{-1} in March and August (Gibbons, 2001). From this analysis it is very evident that at Langtry the Pecos River contained many trace elements. Water hardness was moderate and nutrient chemicals were relatively low.

General Remarks as to Water Quality

Water quality is relatively natural in the upper river, but it can be poor downstream in the lower reaches of the river. Total dissolved solids can be as high as 20,000 mg L^{-1} during low flows in the lower reaches. High TDS is contributed from springs below Colonias and in the Malaga Bend area and by groundwater accretions in the Acme–Bitter Lakes and Lake Arthur areas.

Better water quality is contributed by tributaries and flood flows. Irrigation that occurs in the middle and lower reaches combined with oilfield brines from now unused disposal pits near Artesia may add to the total dissolved solids load. The potash industry in the Carlsbad area disposed of large quantities of salt into unlined lagoons. It is unsure how much of the potash waste or oilfield brine contributed to the river's salinity.

The upper Pecos (in Texas) flows through the Toyah Basin from Red Bluff Reservoir to a point about 35 km south of Girvin. The sediments are Quaternary alluvium of unconsolidated, consolidated, and partially consolidated sand, silt, caliche, gypsum, clay, gravel, and boulders. The meandering river has long pools formed by gravel bars, rock outcrops, and low diversion dams. The river banks support dense growths of salt cedar (*Tamarix* spp.).

The water in the Pecos is very saline by the time it enters Texas. An experimental Salinity Alleviation Project was started in the Malaga Bend area in August 1963. The Pecos receives better water from the Delaware, but the influx is small. The Delaware water is very hard and dominated by calcium sulfate.

Conductivity, an indirect measure of dissolved solids, increased from Orla (12,600 µS cm^{-1}) to Girvin (16,400 µS cm^{-1}), then decreased to Shell Slab (7350 µS cm^{-1}). The largest drop in salinity occurred between Sheffield (14,100 µS cm^{-1}) and Shell Slab (7350 µS cm^{-1}). This dilution is believed to be caused by freshwater inflows from sources in the Cretaceous limestone formations of the Edwards–Stockton Plateau, which the river traverses between Sheffield and Shell Slab. Fresh water is also added to the river by Independence Creek, which is located approximately 15 river miles upstream of Shell Slab. The daily average flow of Independence Creek was estimated between 1975 and 1984 to be 28.4 ft^3/sec (TNRCC, 1996a).

Concentrations of chloride and sulfate anions exhibited the same trend as conductivity. The total dissolved solids calculated were slightly higher at Sheffield than at Girvin. Other sample analyses included an increase in alkalinity from Girvin to Shell Slab. This is believed to be due to the dissolution of Cretaceous limestone formations (Davis, 1987) which the river encounters in the Sheffield area. Also, slightly elevated values for ammonia and nitrate at Sheffield and elevated nitrate levels at Shell Slab reflect a trend toward increased ranching land use in the lower reaches of the river. It was noted that spring flow, especially from limestone formations, frequently contains rather high nitrate concentrations (TNRCC, 1996a; Wetzel, 1975).

ECOSYSTEM DYNAMICS

New Mexico: Aquatic Life

Insects. In a study of the insect fauna at Puerta de Luna and Artesia, New Mexico (Table 4.4) we see a somewhat different distribution than at the stations between Orla and Shumla. The mayfly *Tricorythodes* was most common at Puerta de Luna. Similarly, the Coleoptera were more common at Puerta de Luna than at Artesia. This was also true for the Hydropsychidae, which were much more numerous at Puerta de Luna than at Artesia. This was also true for the trichopteran *Nectopsyche gracilis*, whereas the dipterans *Calopsectra* spp. were much more common at Artesia. Blackflies were far more common at Puerta de Luna than were the *Polypedilum*. As to taxonomic richness and total biomass, the insects were far greater at Puerta de Luna than at Artesia (Table 4.4). In

TABLE 4.4. *Mean Macroinvertebrate Densities (number/m^{-2}) in the Middle Pecos River, New Mexico, July–August 1979*

Site:	Puerta de Luna		Artesia	
Sampler[a]:	Basket	Circular	Basket	Multiplate
Odonata				
Ischneura sp.	6	0	28	65
Other Odonata	6	8	0	0
Ephemeroptera				
Tricorythodes sp.	85	17	0	4
Other Ephemeroptera	29	8	6	0
Coleoptera				
Total	23	0	0	0
Trichoptera				
Hydropsychidae	567	4	28	0
Hydroptilidae	23	0	0	0
Nectopsyche gracilis	130	4	0	0
Diptera				
Calopsectra spp.	0	0	181	0
Chironomus sp.	289	0	0	0
Polypedilum spp.	170	13	0	0
Other Chironomidae	17	4	0	0
Simulium sp.	1562	30	34	0
Hemerodromia spp.	85	4	0	0
Other Diptera	23	0	0	0
Other				
Total other	6	0	28	4
Total density	3021	96	333	102
Total biomass (mg m^{2})	7765	360	435	341
Taxonomic richness	21	12	8	4

[a]Basket sampler, wire cage with concrete spheres; circular sampler, modified Hess sampler; multiplate sampler, modified Hester-Dendy sampler.

1979 the most numerous insects collected at Puerta de Luna were the Hydropsychidae, *Nectopsyche gracilis*, the chironomid *Chironomus* sp., and the blackflies *Simulium* sp. In general, the insects were much more common at Puerta de Luna than at Artesia. No very large numbers of any one species were found at Artesia. The most common species at Artesia were the dipterans *Calopsectra* spp.

Fish Fauna. A study was made of the fish in the middle section of the Pecos River in New Mexico, which extends 346 km from Lake Sumner to Lake McMillan. This area has low relief and the river meanders in a broad valley. The stream has a low gradient and its bed is sand and silt. Thirty-one species of fish were collected. A segment with low relief extended from about 10 km below Lake McMillan. The stream in this reach has a stair-step morphology with alternations between riffles (cobble–gravel–boulders) and deep sand/silt pools. The fish fauna was similar to that of the lower portion of this middle reach of the river (Hatch et al., 1985).

In this study *Notropis simus pecosensis* was less abundant and less widely distributed than in the past. Historically, there were populations from Santa Rosa to Carlsbad. However, in this study this fish was found only in the middle section from Fort Sumner to Artesia. The ability of this species to survive in this area is doubtful. This reach is dewatered periodically when flow regimes are regulated in response to irrigation demands. *N. simus pecosensis* was abundant only between Hagerman and Artesia and in the seepage zone near Fort Sumner. In other parts of the range, it appears to be dependent on sustained runoff and dispersal of individuals from the areas cited above (Hatch et al., 1985).

N. simus is absent between Santa Rosa and Fort Sumner because of the use of fish toxicants and regulation of flow. Other native species have also been eliminated. Recruitment from downstream allowed *N. simus* to exist in this upper reach. Since the completion of Sumner Dam in 1937 (forming Sumner Lake), there has been a decline in this species. It is probable that this species has not been significant in this reach since the completion of McMillan Dam at McMillan Lake in 1893. Only a few spring outflow zones serve as desirable habitat for this species. *N. simus* appears to have a prolonged spawning season. Spawning probably ends shortly after September. Fecundity seems to be less in this reach for *Notropis girardi* (Hatch et al., 1985).

Thirty-one species were recorded in the historic range of *N. simus* (Santa Rosa to Fort Sumner). Of these, only 23 were recorded within the present range of this species by Hatch et al. (1985). Twenty species were collected with *N. simus*. However, only five could be considered frequent associates: *Hybognathus placitus*, *Hybopsis aestivalis*, *Notropis jemezanus*, *N. lutrensis*, and *N. stramineus*. The first three species are ecologically similar to *N. simus*. *N. simus* was found in every major habitat encountered from Fort Sumner to Artesia, except in stagnant pools. They were collected primarily in the main channel sites, which had sandy substrates, low velocity, laminar flow, and depths of 17 to 41 cm. As the fish matured, they more commonly used the main channel sites. The young-of-the-year were collected in large numbers in backwaters. However, they were equally abundant in the main channel (Hatch et al., 1985).

N. girardi was introduced into the Pecos River. It was probably released as a bait fish by anglers below Sumner Dam in 1978. The source of the original stock of this species is unknown, but there is probably a geographically close native population in the Canadian River drainage, and this fish is probably the source of the specimens that were introduced. This fish dispersed downstream at least 260 km by the autumn of 1981. By 1987, *N. girardi* was verified downstream another 98 km to the New Mexico–Texas border. Multiple introductions may have occurred of *N. girardi* throughout the Pecos River (Bestgen et al., 1989).

It is interesting to note that the flow regimes in the Pecos River, which were established in 1986, mimic the extaordinary variability of the regime formerly present in the native range of *N. girardi*. Annual irrigation demands peak during the spawning season of *N. girardi* and are unpredictable, high, and last for days to weeks, intervals sufficient to allow successful reproduction by *N. girardi* (Bestgen et al., 1989).

The first collections of *N. girardi* were taken from the Pecos River (Pecos, New Mexico) on September 7, 1978 just downstream from Sumner Dam. Collections made from September 1981 to February 1982 indicated that *N. girardi* had become widespread and abundant in the Pecos River. Relative abundance and population stability varied substantially throughout the study area and across time for *N. girardi*. It was found regularly just below Sumner Reservoir downstream to near Carlsbad where it was found in 17 collection sites. The largest collection of *N. girardi* was 472 specimens. In 1986–1987, *N. girardi* was in 21 of 24 collections. It was more common in the 1986–1987 period than in the period 1981–1982. In both 1981–1982 and 1986–1987, *N. girardi* was uncommon immediately below Sumner Reservoir and between McMillan and Red Bluff Reservoirs. This species had at least three and possibly four age classes. The dominant age class was 0 to 1 year. Collections in July and August suggest that spawning peaked between mid-June and mid-July in this stream (Bestgen et al., 1989).

The introduction of *N. girardi* into the Pecos River, probably related to bait-fish anglers below Sumner Dam in 1978. The success of *N. girardi* in the Pecos River was initially perplexing given its diminished stature in its native range. The native fish community of the Pecos River has more species than are generally found sympatric with *N. girardi* in its native range, and competitive interactions with established native species presumably rendered establishment of a nonnative species more difficult. Life history parameters of the introduced Pecos River population and native populations of *N. girardi* in New Mexico and Kansas appear similar. Age growth, age structure, mortality, and population dynamics are especially similar (Bestgen et al., 1989).

Hybognathus placitus, which is an abundant plains streams species, was introduced into the Pecos River. It occurs widely within the habitat of *N. girardi* in its native range. *Hybognathus placitus* co-occurs widely with *N. girardi* in its native range, and because of its size and abundance, the former species is actively sought as bait. It may have been accidentally introduced in a bait shipment. This species, *H. placitus*, is now a most abundant species in the Pecos River, and it comprises more than 80% of the catch at some sites. Its success in the Pecos drainage is probably due to the same factors that have benefitted *N. girardi*. The establishment of *H. placitus* in the Pecos River coincides with the disappearance of the native *Hybognathus amarus*. This species has not been taken from the Pecos River since 1968.

Pecos Gambusia (*Gambusia nobilis*) once occurred throughout the Pecos River drainage. Now it is rated as endangered and found only in seven isolated locations at Bitter Lake National Refuge (Chaves County, New Mexico) and at one locality at Blue Springs (Eddy County, New Mexico). Pecos Gambusia uses submerged cliff overhangs and aquatic vegetation, including *Chara* sp., *Ruppia maritima*, and *Potamogeton pectinatus* for cover. In Blue Springs, however, Pecos Gambusia uses primarily *Paspalum districhum*, *Nasturtium* sp., *Chara* sp., and *Eleocharis* sp. as habitat. Any shallow area with aquatic vegetation seems suitable as long as other factors are within its tolerance limits. No Pecos Gambusia were found where the hardness of the water exceeded 5100 mg L^{-1} or conductivity was greater than 18,000 µS. The distribution of the species seemed to be limited or prohibited by the lack of aquatic vegetation. Pecos Gambusia fed on dipterans and other surface or subsurface insects of suitable size. They preferred to feed at night, but they could feed in the daytime as well.

Coexisting with Pecos Gambusia were an unidentified species of *Cyprinodon* (Pecos pupfish), *Fundulus zebrinus* (Rio Grande killifish), *Lucania parva* (rainwater killifish), *Notropis lutrensis* (red shiner), and *Dionda episcopa* (roundnosed minnow). Larger populations of Pecos Gambusia were correlated with small spring flows and seepages along the main spring run. Pecos Gambusia can compete effectively with *Gambusia affinis* in certain environmentally stable habitats. They occurred together only in spring-fed waters. *G. affinis* has a higher rate of reproduction than *G. nobilis*, giving *G. affinis* an advantage in unstable environments. Pecos Gambusia comprised approximately 28% of the total Gambusia population at Blue Springs (Bednarz, 1979).

Texas: Aquatic Life
In the lower reach, two stations were located. One was at U.S. Highway 290 (Figure 4.2) southeast of Sheffield at river km 417 in a riffle run on a streambed of calcite and coarse gravel. Periphyton, mainly diatoms, usually covered the substratum, and the immediate area was devoid of rooted aquatic plants. The other station in this lower reach of the river was at Shumla in a riffle-run area on gravel and sand substratum. Sparse growth of the green filamentous alga *Zygnema* and the attached diatom *Cymbella* were usually present on the substratum. The shoreline vegetation and aquatic macrophytes were absent at the beginning of the study at Shumla due to scouring by a severe flood in September 1974. However, by 1977, dense stands of cane grass and small numbers of salt cedar saplings were present on the banks. Dense beds of *Chara* and *Potamogeton* were growing in shallow pools and backwaters where recently deposited finer sediments had accumulated (Davis, 1980).

According to Rhodes and Hubbs (1992), a fish kill by the golden alga *Prymnesium parvum* occurred in the Pecos River since 1985. A similar fish kill was also reported in the 1960s. These blooms have resulted in substantial fish kills. The extent has often reached Amistad Reservoir. There was a massive fish kill in November 1986. Only six species—*Dionda episcopa*, *Cyprinella lutrensis*, *Cyprinodon* hybrids, *Lucania parva*, *Fundulus grandis*, and *Gambusia affinis*—were able to survive and were collected two months following the bloom. These species were either salt-tolerant species such as *L. parva* or less tolerant species found in the extreme lower river (e.g., *C. lutrensis* and

D. episcopa). In a bloom in an adjacent spring the following year, *Cyprinodon* appeared to be fairly tolerant to the algal toxin (Rhodes and Hubbs, 1992).

Macroinvertebrates. Invertebrate studies were made in the Pecos River northeast of Orla (FM 652); north of Coyanosa (FM 1776); east of Girvin (U.S. 67); east of Sheffield (U.S. 290); and just upstream of Val Verde County at a location known as Shell Slab (Figure 4.2).

The benthic data were divided into three broad categories: community indicators, prevalence of sensitive taxa, and functional feeding groups (TNRCC, 1996a). These parameters were used to provide a quantitative means for assessing benthic macroinvertebrate community structure at each of the sampling stations (TNRCC, 1996a). Each of the habitats were analyzed as to riffles, pool depth, bank stability, riparian cover, flow fluctuations, channel sinuosity, bottom substrate, and aesthetics (TNRCC, 1996a).

As seen from Table 4.5, the macroinvertebrates were quite diverse in the Pecos River. The oligochaetes were most numerous at Orla in the period between 1976 and 1977. The ostracods were most numerous at Girvin.

Stream Communities. The river after leaving Red Bluff Reservoir travels approximately 14 miles before reaching FM 652 northeast of Orla.

Orla Just Upstream of the FM 652 Bridge. The benthic macroinvertebrate community was limited in taxa due to the poor habitats and relatively high salinity. The number of invertebrate genera was only eight, and this site had the fewest number of individuals captured. Macroinvertebrates were also found in relative abundance at Coyanosa and Girvin. The most saline stretch of the river was between Orla and Girvin. In this reach the genera *Berosus* sp. and *Argia* sp. were well established. There were no representatives of intolerant species such as those belonging to the mayflies, stoneflies, or trichopteran orders. The EPT (Ephemeroptera–Plecoptera–Trichoptera) index was zero. The trophic composition of the river was imbalanced because of the dominance of gatherers, including *Palaemonetes pugio* and *Berosus* sp. (TNRCC, 1996a).

The macroinvertebrate community northeast of Orla (FM 652) was composed of the decapod *Palaemonetes pugio*, which represented 18.6% of the invertebrate fauna. Eight specimens were collected. Several Coleoptera were also collected. They belong to the genus *Berosus* sp. This taxon formed 46.5% of the community. Also collected were *Paracymus subcupreus*, which formed 2.3% of the community and has a tolerance to pollution of 4. *Tropisternus lateralis* (2.3% of the population) has a tolerance of 9.8. All three of these taxons were gatherers. Another beetle collected was *Pronoterus* sp. It has a tolerance of 6.9. It is a predator. Four specimens of the dipteran, *Dicrotendipes neomodestus*, a chironomid, were collected. They formed 9.3% of the community and had a tolerance of 8.3. It is a gatherer, filterer, and scraper. Of the damselfly, *Argia* sp., five specimens were collected. It formed 11.7% of the population. Its tolerance is 8.7 and it is a predator. Of the gastropod (snail) *Physella virgata*, three were collected. They formed 7% of the population. Its tolerance was 8 and it is a scraper (TNRCC, 1996a).

North of Coyanosa (FM 1776). North of Coyanosa (FM 1776) the flow is much less than at Orla. This is caused by extensive withdrawals from the river via irrigation canals between Mentone and this site (Grozier et al., 1968). There seemed to be a considerable loss in flow at this station, due to evaporation and transpiration. Bottom substrate and channel sinuosity were poor at this site. However, the habitat score was higher than at Orla, due to more moderate flow fluctuations. The channel was relatively narrow and the river meandered through several bends, where it frequently opened into short pools. In the areas where velocity was relatively fast, the substrate was scoured and sandy. However, the majority of the streambed consisted of a silty-clay layer overlying a deep, black, anaerobic organic layer (TNRCC, 1996a). Macrophytic growth is abundant in certain places in the area; however, it is not as dense as at Girvin. In this area dense stream-side salt cedar growth occasionally hung into the water. The left bank is fairly steep where the river runs along a small rocky bluff outcrop. Portions of the right bank are very flat, allowing for overflow during high-flow conditions (TNRCC, 1996a).

In general, in this area north of Coyanosa, flow fluctuations were moderate and channel sinuosity was low. The bottom substrate was composed of sand, silt, clay, or bedrock. It was relatively unstable. Along the banks of the river were trees or native vegetation. The water was "discolored" (TNRCC, 1996a).

The macroinvertebrate community in the river at Coyanosa indicated saline conditions. Next to Girvin, Coyanosa was the second-most saline site in the study. Although salinity is higher than at Orla, habitats improved to the point that the site could support greater benthic macroinvertebrate community integrity. This site had the second highest number of individuals, 327. Sixty-three percent of these were the amphipod, *Hyalella azteca*. This species is commonly found in brackish water (Pennak, 1978). Other taxa present in abundance were *Berosus* sp. and *Argia* sp. Four species of chironomids were also found in this area, as were two Trichoptera genera, *Cheumatopsyche* and *Ithytrichia*, and an Anisoptera genus, *Erpetogomphus*. All of these taxons have relatively low pollution tolerance. Most of the insects in this community were gatherers, including an abundant population of *Hyalella azteca*. Decaying vegetation formed a plentiful source of food, as macrophytes were entering dormancy at the time of sampling. Predators were well represented. They included the Odonata taxa, *Erpetogomphus* sp. and *Argia* sp. A total of 327 individuals were collected, representing 12 genera. Of these, 206 were the amphipod *Hyalella azteca*; 13 were the beetle *Berosus* sp.; specimens of four taxons of Diptera were collected (*Cricotopus bicinctus*, 1; *Dicrotendipes neomodestus*, 15; *Polypedilum digitifer*, 1; and *Tanytarsus glabresens* gr., 14); the hemipteran *Lethocerus* sp., 1 specimen; the dragonfly *Erpetogomphus* sp., 10 specimens; and the damselfly *Argia* sp., 49 specimens. Other insects collected were the caddisflies (11) [*Cheumatopsyche* sp. (10) and *Ithytrichia* sp. (1)]. The gastropod, a snail *Physella virgata*, was represented by six specimens. The largest population collected was amphipods, 206 individuals, which comprised 63% of the population. Their tolerance rating was 7.9. The next most common group was the damselflies (Zygoptera), 49 specimens. The Zygoptera had a slightly higher tolerance rating than the amphipods. They were predators, whereas the Amphipoda *Hyalella azteca* was a gatherer. These invertebrates comprised the community at Coyanosa (TNRCC, 1996a).

East of Girvin on U.S. 67.   In the Pecos River east of Girvin on U.S. 67, there was a considerable amount of instream cover. The structure of the river was pools (2 to 4 ft deep) and riffles that were relatively rare. The banks were moderately stable.

TABLE 4.5. *Relative Macroinvertebrate Densities (%) for Suber Sampler Collections at Four Texas Stations in the Pecos River, 1976–1977*

Station: River km from Mouth:	Orla km 602	Girvin km 314	Sheffield km 212	Shumla km 30
Oligochaeta				
Total	16.6	5.1	5.9	0.6
Ostracoda				
Total	0	20.9	0	0.1
Odonata				
Argia sp.	2.1	0.1	4.4	0.2
Other Odonata	0.8	0.5	<0.1	1.1
Ephemeroptera				
Choroterpes mexicanus	0	0	7.6	4.6
Thraulodes sp.	0	0	0	19.7
Traverella sp.	0	0	0	6.4
Tricorythodes sp.	0	0	0	14.8
Other Ephemeroptera	0	0	0	0.6
Hemiptera				
Naucoridae	0	<0.1	0	7.2
Other Hemiptera	0	<0.1	0	0
Coleoptera				
Elmidae	0	0	<0.1	12.1
Berosus spp.	0.9	10.9	0.4	0
Other Coleoptera	0	0	0	2.4
Trichoptera				
Cheumatopsyche spp.	26.2	0.2	19.8	6.7
Hydroptilidae	7.3	<0.1	8.7	5.5
Other Trichoptera	0	0	0	5.0
Diptera				
Cricotopus bicinctus	1.9	<0.1	18.3	1.5
Dicrotendipes neomodestus	6.8	0.2	3.7	0
Pseudochironomus sp.	9.2	0	27.0	0.1
Tanytarsus spp.	21.5	55.6	1.6	1.2
Other Chironomidae	4.8	3.8	0.7	3.9
Other Diptera	1.6	0.1	0.1	1.1
Other				
Noninsect	0	23.5	1.5	1.9
Insect	0	2.9	0.7	24.3
Taxonomic richness	17	21	24	50

Source: Modified from Davis (1980).

However, there was some evidence of erosion. The riparian cover was moderate. Flow fluctuations were moderate and there was evidence of debris along the middle portion of the banks. Channel sinuosity was low. The bottom substrate was unstable in various areas. It was composed of sand, clay, silt, or bedrock. Along the banks, trees and native vegetation were common. The water was "discolored" (TNRCC, 1996a).

Because of rather high salinity at this point (16,400 µS), the taxa were salt tolerant. They were *Hyalella azteca*, the dominant genus, comprising 53.9% of the community; and *Berosus* sp. and *Argia* sp., which were relatively abundant. This site had the highest biotic index in the study, 8.11 (TNRCC, 1996a).

The dominant species here was the amphipod, *Hyalella azteca*; 152 specimens were collected of this species. Other abundant species were the beetle *Berosus* sp. (69 specimens) and the damselfly *Argia* sp. (17 specimens). The mayfly *Callibaetis* sp. (1 specimen) was collected. This taxon is usually found in waters of fairly good quality. However, it has a high tolerance value of 9.3. Found in high numbers was the gastropod *Physella virgata* (30 individuals), which indicates abundant periphytic growth on the boulder substrates. The community trophic structure was dominated by gatherers, primarily *Hyalella azteca* (the amphipod) and *Berosus* sp. (a beetle), which together comprised 75% of the community. Decaying macrophytes were abundant and potentially an important food source on which *Hyalella azteca* was feeding. The gastropod, *Physella virgata*, is a scraper. Thirty specimens of this species were collected. It formed 10.6% of the community. It is clear that the community was dominated by amphipods, followed by the beetle *Berosus* sp., the gastropod *Physella virgata*, and the damselfly *Argia* sp. The rest of the species had small populations that were no greater than 10% of the specimens collected (TNRCC, 1996a).

East of Sheffield at U.S. 290.   In this region the river flows into a geological area characterized by Cretaceous limestone formations. Historically, the flow increased in this area because of hydraulic communication with springs and groundwater (Davis, 1987). Although the bed of the river is deep and silty in places, there is a noticeable trend toward sand and gravel as the substrate in many areas. Riffles were encountered with more frequency than upstream. Also, bends in the river created deeper pools, which were attractive fish habitats (TNRCC, 1996a).

Around a curve the stream velocity increases and the channel narrows. Most of the substrate within the riffle consists of large cobbles and boulders. The maximum pool depth is greater than 4 ft. Some evidence of erosion is present on the banks. The side slopes are up to 40° on one bank. The natural cover varies from 15 to 150 ft. The bottom substrate is 10 to 30% gravel or larger substrates. Trees and native vegetation are common, and it should be noted that the stream water was somewhat discolored.

At Sheffield the inflow of fresh water from Cretaceous limestone formations varies, as it is recorded in the literature as being large but was not large at the time of these studies (September 20, 1994). Only 55 invertebrates were captured. Most of these were *Physella virgata*, which comprised 49.1% of the population. Eleven genera were collected; six were chironomids of varying pollution tolerance (Table 4.7). Similarly to the situation in the three saline sites upstream, *Hyalella azteca* and *Berosus* sp. were present. The community structure consisted of almost equal representations of gatherers, including *Hyalella azteca*, some coleopterans (beetles), a few chironomid taxa, and scrapers, dominated by *Physella virgata*. These two trophic groups (scrapers and gatherers) comprised about 90% of the community, indicating that primary food sources were decaying vegetation and periphytic growth on the substrate (TNRCC, 1996a).

It is unclear why the benthos community was not well developed. Toxic impact is unlikely, as there is no industrial activity in the area. This lack of development may be due to the speed of the current passing through the boulder-dominated riffle, preventing sufficient attachment to support a thriving community, due to continuous scouring. Historically, the site has supported a prolific benthic macroinvertebrate community. However, this was not apparent at this time (TNRCC, 1996a).

In the Pecos River East of Sheffield (U.S. 290) the benthic community was dominated by the gastropod (snail) *Physella virgata*, which formed 49.1% of the community. The other most common group were the true flies. The amphipod *Hyalella azteca* was represented by six specimens, and the beetles *Berosus* sp. and *Lutrochus luteus* were represented by two specimens each. The Diptera (true flies) were *Chironomus riparius* gr. (one specimen), *Dicrotendipes neomodestus* (one specimen), *Goeldichironomus natans* gr. (seven specimens), *Pseudochironomus* sp. (six specimens), *Telopelopia okoboji* (one specimen), and *Tabanus* sp. (one specimen). Of the hemipteran bug *Cryphocricos hungerfordi*, one specimen was collected. Its tolerance was 4 and it is a predator. Of the gastropod *Physella virgata*, 27 were collected. It formed 49.1% of the population. It has a tolerance rating of 8 and it is a scraper. The total macroinvertebrate fauna was 55, total genera 11 (TNRCC, 1996a).

Shell Slab.   Shell Slab had a much better habitat, due to increased flow and predominance of gravel/cobble/boulder substrate. The increased flow was caused by inflow from springs and groundwater seepage, as well as the convergence of Independence Creek approximately 15 mi upstream. In this area the stream channel was broader than in the upper part of the Pecos River (in Texas). This created abundant riffle habitat. Bends provided structure in the form of deep pools and undercut banks. The stream banks were gradually sloping and to some extent covered by riparian salt cedar. However, this was less dense than upstream, as the proportion of live oak is greater here (TNRCC, 1996a).

The sampling was done largely on riffles that supported a variety of substrates exclusive of fine silt. Some undercut banks with vegetation/roots also were substrates that were sampled. Riffles were common, pools were deep, and the banks in this area were stable. The riparian cover was moderate, and flow fluctuations were moderate. The channel was quite sinuous, and the bottom substrate was stable, composed primarily of cobbles, rubble, or gravel. The watershed was a natural wooded unpastured area (TNRCC, 1996a).

Twenty-eight macroinvertebrate genera were collected and more individuals were captured (545 total) than at any of the earlier sites. The dominant genus and species was *Macrelmis texana*, which comprised 27.7% of the community. The commonness of this species reflects a stable community with an even distribution of individuals among many taxa. Numbers of scrapers and filter feeders were approximately equal, reflecting a balanced food source of periphyton and suspended organic matter. The site exhibited a low biotic index value, indicating that the benthic macroinvertebrate community as a whole is relatively pollution intolerant. Ten EPT (Ephemeroptera–Plecoptera–Trichoptera) taxa were collected, five times more than at the site with the next-highest number, which was Coyanosa. Aside from the gatherer feeding group dominated by coleopterans, which comprised about half of the community, there was almost equal representation by scrapers, filter feeders, and predators, a further indication of community stability. Certain taxa common at more saline sites upstream were absent here. Examples were *Berosus* sp. and *Hyalella azteca*. They evidently are unable to compete under local environmental conditions that allow intolerant taxa to proliferate (TNRCC, 1996a).

Beetles that were present were *Helichus* sp. (1), *Hexacylloepus ferrugineus* (1), *Macrelmis texana* (151), *Microcylloepus pusillus* (3), *Stenelmis* sp. (33), *Lutrochus luteus* (12), and *Psephenus texanus* (1). *Macrelmis texana* formed 27.7% of the population and was by far the most common species. The Diptera were *Rheotanytarsus exiguus* gr. (1), *Simulium* sp. (14), and *Tabanus* sp. (4). The mayflies (Ephemeroptera) were *Baetis* sp. (8, or 1.5% of the total; gatherer/scrapers); *Camelobaetidius* sp. (8, or 1.5% of the total; gatherer/scrapers); *Caenis* sp. (1, or 0.2% of the total; gatherer/scrapers); *Neochoroterpes* sp. (124, or 22.7% of the total; gatherer/ scrapers); and *Tricorythodes* sp. (5, or 0.9% of the total; gatherers). Of the Hemiptera (bugs), *Ambrysus* sp., 24 were collected (4.4% of the total) and of *Cryphocricos hungerfordi* 50 were collected (9.2% of the total). They are predators. Of the Megaloptera (hellgrammites) *Corydalus cornutus*, 2 were taken (0.4% of the total). They are predators. Of the dragonflies (Anisoptera), there were *Phyllogomphoides albrighti* (1, or 0.2% of the total), and immature Gomphidae (immature dragonflies) (2, or 0.4% of the total). They are predators. The Zygoptera or damselflies were represented by 3 specimens of *Argia* sp. (0.5% of the total). It is a predator. The Trichoptera (caddisflies) were represented by *Cheumatopsyche* sp. (36, or 6.6% of the total), *Hydropsyche* sp. (2 or 0.4% of the total), *Smicridea* sp. (4, or 0.7% of the total), *Chimarra* sp. (31, or 5.7% of the total), and *Dolophilodes* sp. (2, or 0.4% of the total). All of the Trichoptera were filter feeders.

Of the Gastropoda (snails), *Physella virgata*, 20 were collected or 3.7% of the total. It is a scraper. Of the Pelecypoda (clams) *Corbicula fluminea*, 1, or 0.2% of the community, were collected. It was a filter feeder. The total number of macroinvertebrates collected was 545, representing 28 genera. Thus it is evident that the diversity here is that of a natural community (TNRCC, 1996b).

Water Quality and the Pecos River Macroinvertebrates in Texas.   The invertebrate fauna was poorest at the FM 652 bridge site. This was probably due to a large extent to flow fluctuations, caused by varying demands by irrigation districts. The streambed was scoured, which provided poor habitats for macroinvertebrates. The water was fairly saline (12,600 µS conductivity), and the predominant organisms were *Berosus* sp. and *Argia* sp. Taxa richness (8) was the lowest in the study area. No intolerant Ephemeroptera–Plecoptera–Trichoptera taxa were represented (TNRCC, 1996a).

In the river in the Coyanosa area, between Orla and Girvin, water quality was degraded by high conductivity due to evaporation and transpiration as well as continued dissolution of Permian salts. However, the habitats were fairly good and the aquatic system could support more diverse benthic macroinvertebrates. Here there were 12 taxa. Although the community was considered salt tolerant, two relatively intolerant trichopteran genera were present: *Cheumatopsyche* and *Ithytrichia*. The dominant genus, however, was the salt-tolerant amphipod, *Hyalella azteca*. Other taxa tolerant of salt water included *Berosus* sp. and *Argia* sp. (TNRCC, 1996a).

The salinity was highest at Girvin (16,400 µS conductivity). Although habitat was better than at Coyanosa, extreme salinity suppressed the benthic macroinvertebrate community. Taxa richness was minimal (8), and the biotic index value was the highest in the study (8.11). The dominant species was *Hyalella azteca*, which comprised 53.9% of the community. Other taxa included *Berosus* sp. and *Argia* sp. (TNRCC, 1996a).

The Sheffield site was an anomaly. Although water quality (14,100 µS conductivity) and habitat were better there than they were at Girvin, the benthic macroinvertebrate community did not improve to the degree that might have been expected. Only the Orla site had fewer individuals captured. The dominant genus was the gastropod, *Physella virgata*. Historically, the site has supported an abundance of benthic macroinvertebrates, but due to high salinity in the river at the time of the study it did not (TNRCC, 1996a).

Finally, as the river approaches Val Verde County, at Shell Slab site, the water was diluted by groundwater and inflow from Independence Creek. In general, the physical habitat was exceptional, as Shell Slab scored highest among sites in almost all categories of the habitat quality index. As a result of reduced salinity and favorable habitat, the site supported the richest benthic macroinvertebrate community in the study. All tests associated with biotic integrity reflected optimal environmental conditions at Shell Slab (TNRCC, 1996a).

Comparison of Orla, Girvin, Sheffield, and Shumla. It will be noted that the total oligochaetes were much more numerous at Orla than at the other stations, whereas the ostracods were much more numerous at Girvin. The Ephemeroptera were much more common at Shumla than at the other stations (Table 4.5). This was also true for the Naucoridae, although they were never very common. Elmids were also much more common at Shumla, whereas the unidentified species of *Berosus* were more common at Girvin. Cheumatopsyche were much more common at Orla and Sheffield than at the other stations.

Cricotopus bicinctus was most common at Sheffield. *Pseudochironomus* were much more common at Sheffield, whereas *Tanytarsus* was much more common at Girvin, followed by Orla than at the other stations. As one might expect, the greatest number of species was at Shumla, followed by Sheffield, with the fewest species at Orla (Table 4.5).

The most common dipterans at Girvin were the *Tanytarsus* (Chironomidae), which formed 55.6% of the fauna collected. The next greatest abundance of the Chironomidae was at Orla, where the Chironomidae other than *Tanytarsus* seemed to be most common (Table 4.5).

Taxonomic richness (number of species in a given number of individuals) was similar at Orla, Girvin, and Sheffield but was almost double at Shumla. In general, the macroinvertebrate taxonomic richness increased downstream from Orla (20 species) to Girvin (26 species) to Sheffield (32), to Shumla (54). The differences are associated with water quality, primarily salinity and total dissolved solids. Similar longitudinal patterns of diversity were evident (Davis, 1980).

Variations in occurrence of invertebrates at Orla were associated with flow regulation at Red Bluff. Variations were low during the irrigation season, when irregular releases resulted in scouring and sudden salinity fluctuations. Winter maximums were associated with long periods of no releases, stable low flows, and stable, albeit high, salinities.

Variation in occurrence at Girvin was similar in autumn and winter for each year. But spring and summer values varied from year to year. Variation appears to depend on autotrophic standing crops that were roughly associated with seasonality. Standing crops of spermatophytes generally resulted in high taxonomic richness and high densities of invertebrates. The associations were dominated by a few extremely abundant taxa. Spring and summer variability of species seemed to be caused by the prevalence of detritivores and irregularity in flow. The significant differences in the amount of detritus in the river may significantly influence the number and kind of detritivores (Davis, 1980).

Seasonal Diversity of Insects.   At Orla, Girvin, and Sheffield the seasonal diversity of insects seemed to be caused more by one or more taxa establishing or losing dominance than by a change in the size of populations of most of the species. Conversely, seasonal changes in diversity at Shumla were induced to a greater degree by changes in the number of taxa than by changes in dominance, as d varied directly with the number of taxa, and d and r were more independent (Davis, 1980).

Seasonal diversity at all stations had no direct correlation to seasonal physico-chemical changes, which varied irregularly. Diversity of species at Orla varied directly with the duration of constant-volume releases from Red Bluff Reservoir prior to sampling (Davis, 1980).

Diversity was usually low at Orla during the irrigation season when reservoir releases were irregular, resulting in sudden salinity fluctuations and scouring of the substratum. This condition apparently favored the few asymptotically adaptable taxa, which become dominant (Davis, 1980).

The insect fauna of the Pecos River is diverse but apparently has not been as well studied as the fish fauna. The mayfly (Ephemeroptera) Heptageniidae was present. Coleoptera were represented by the Elmidae. These are the common beetles, and they are widespread. They feed on plant material and can be found on wood, in gravel, or in rocks in streams throughout the area. Six taxons of Elmids were found in the Pecos River. Various Diptera have been collected in the Pecos River. The larvae of the Rhagionidae inhabit the Pecos River. Specimens of Rhagionidae have been collected from the Pecos River. Trichoptera, commonly known as caddisflies, were found in both running water and in pools. They feed mainly on plant materials. The Hydropsychidae were represented by the caddisflies. They were filter feeders. Other insects found in the Pecos River were Chironomidae (Diptera). Taxons of this group were found in many different habitats. They may be carnivores or herbivores. Thirty-six specimens have been collected in the Pecos River. Tipulids, which are Diptera, live in the damp soil, moss, and rotten wood or decaying leaves. Some of them prey on other insect larvae, while others are herbivorous and feed on roots of grass. Stoneflies (Plecoptera) were found in the Pecos River. Seventy-five Perlidae nymphs

(Plecoptera) have been collected in the Pecos River. These species take about three years to develop. They prefer cooler water. Other Plecoptera are the Taeniopterigidae, which are mainly herbivores feeding on decaying leaves, algae, and detritus (Hunt, 1997).

Comments: Fish Fauna of the Pecos River in Texas. *Cyprinodon pecosensis* was found near Girvin and also near Sheffield. The bulk of its diet is diatoms and detritus. However, it is an omnivore. Pupfishes typically feed as omnivores. Other pupfishes found were *Cyprinodon atrorus, C. variegatus, C. macularius, C. rubrofluviatilis, C. diabolis, C. bovinus,* and *C. nevadensis.* Animal material was also ingested and seemed to be second in volume. However, it was third in frequency of occurrence. Filamentous algae was third in volume and second in occurrence. Bottom foraging of the fish was indicated by the presence of sand in 50% of the fish. It appears that dietary differences among individual fish may be caused by the habitats of the sexes. Males ingested 25 food items versus 19 in females. Five of the six items were unique to males, and the males ate more harpacticoid copepods, resulting in a high incidence of parasitism by *Acanthocephala.* This parasitism occurred 2.8 times as frequently in males as in females. The more aggressive behavior of males and their having a higher degree of territoriality may be the cause of the differences in amounts consumed between males and females. Animal items were more voluminous in fish with gut lengths of less than 50 mm. The importance of gut length seemed to decline in larger fish. Filamentous algae and macrophytes increased in importance as fish grew [similar to what was found in *C. nevadensis amargosae* and *C. bifasciatus*]. It is interesting to note that the number of animal taxa and the volume of animal material ingested by the pupfish varied inversely with the chloride content of the water.

Genus *Cyprinodon* species were remarkably plastic, being euryhaline and eurythermal, traits that could adapt to harsh desert environments (Liu, 1969). Because the pupfishes have a high degree of dietary adaptability, they are able to live on whatever food exists, although the food available may be less desirable and marginally nutritious (Davis, 1981).

The Pecos pupfish is a genetic introgression of an ancestral population of *Cyprinodon rubrofluviatilis* and an earlier resident, possibly *C. bovinus* of the lower Pecos. The new species probably has been isolated from *C. rubrofluviatilis* since before the Wisconsin tide, and individuals of the two species are easily distinguishable. *Cyprinodon pecosensis* was found in the Pecos River. *C. pecosensis* is probably the fusion of an ancestral population similar to *C. rubrofluviatilis* and an earlier resident, *C. bovinus,* and this has resulted in *Cyprinodon pecosensis* (Echelle and Echelle, 1978).

Cyprinodon pecosensis occurs in the Pecos drainage from the Roswell, Chaves County, New Mexico area to the mouth of Independence Creek (Sheffield, Terrell County, Texas). It occurs in a wide variety of habitats, from saline springs and gypsum sinkholes near Roswell to typical desert streams with great fluctuations in flow and physicochemical parameters; although collected in low salinity, *C. pecosensis* is most typically abundant in high-saline habitats that support few species. The most common fish associated with *C. pecosensis* are *Fundulus zebrinus, Lucania parva, Gambusia affinis,* and *Notropis lutrensis.* They are also commonly seen with *Gambusia nobilis* and *Etheostoma lepidum. C. pecosensis* also has dense populations in gypsum sinkholes devoid of *Lepomis cyanellus*, but is sparse in sinkholes containing the latter species (Echelle, 1975; Echelle and Echelle, 1978).

C. elegans and *C. bovinus* are also endemic to the Pecos system. They occur allopatrically in widely separated spring-fed watercourses in the Toyah Creek and Leon Creek drainages of the lower Pecos. None of these three Pecos River pupfishes have been collected in waterways supporting one of the others. There may be a history of ecological isolation in the lower Pecos River with *C. elegans* in fresh water and *C. bovinus* in more saline habitats. Ecologically, *C. pecosensis* is more similar to *C. bovinus* than to *C. elegans*. Both *C. pecosensis* and *C. bovinus* occur abundantly in saline stenothermal spring outflows and in more variable habitats. *C. elegans* is sometimes characterized as a fast-water species compared with other pupfishes (Itzkowitz, 1969; Stevenson and Buchanan, 1973) and is known only from relatively fresh waters. *Cyprinodon* species interbreed with ease (Echelle and Echelle, 1978).

One of the sites collected in the Pecos River in Texas was Girvin. Here the water was clear and mucky substrate supported patches of filamentous algae. Widgeon grass and *Chara* were present upstream. Common in this habitat were the fish *Cyprinodon pecosensis*, *Fundulus zebrinus*, *Lucania parva*, and *Gambusia affinis* (Davis, 1981). Sheffield was a similar site and supported the fish fauna noted in the description of the ecosystem at Sheffield (pages 199–200): *Cyprinodon pecosensis*, *Notropis lutrensis*, *Gambusia affinis*, *Lucania parva* (numerous), *Dionda episcopa*, *Fundulus zebrinus*, and *Menidia beryllina*. Juvenile *Cichlasoma cyanoguttatum* and *Lepomis cyanellus* were less common (Davis, 1981).

Although the diet of these fish was largely detritus and diatoms, Entromostracans were secondarily important and filamentous algae were moderately common. The difference in preference of animal prey appears to be related to the prey abundance, since there were relatively few macroinvertebrates and entromostracans in the Pecos River. However, more were collected in the tributaries. The important algal food of these fish in the river appeared to be the diatoms, *Nitzschia*, *Cyclotella*, *Synedra*, *Amphora* and *Rhizoclonium*. The blue-green algae *Oscillatoria* was also found in the guts (Davis, 1981).

Hybognathus amarus formerly was distributed from New Mexico in the Rio Grande and from the Pecos basin to the Gulf of Mexico. Evidence from collections suggests that this was one of the most abundant species in the basin. However, it has not been reported in recent times from many portions of its former range. Extirpation was probably caused by hybridization and competition with *H. placitus*. It is interesting to note that *H. placitus* was introduced as early as 1968 and *H. amarus* was last recorded in the Pecos in 1968. These facts suggest that competition with *H. amarus* probably occurred. It is interesting to note the correlation between the number of taxons and salinity. Twenty-one taxons were reported from the middle section, 15 from the saline section, and 29 species from the lower section of the Pecos River, which was fresher.

Several Gulf Slope species are otherwise absent from the Rio Grande drainage. These include *Etheostoma lepidum*, *Phenacobius mirabilis*, *Catostomus commersoni*, *Fundulus zebrinus*, and *Semotilus atromaculatus*. There seemed to be some headwater exchange between the Pecos and the Rio Grande in fishes during Pleistocene and Pliocene times.

The shift in climate toward aridity during the late Tertiary resulted in considerable extinction of the fauna for much of the Rio Grande system, particularly in the Pecos. For example, the river was formerly productive, and diverse fauna of Salmonids and Centrarchids were much more common and now are greatly reduced.

Several species are reported in the Pecos drainage that are not reported in other studies in the Rio Grande. These species occur in other than mountainous regions. They include *Anguilla rostrata, Dionda episcopa, Gila pandora, Hybognathus amarus, Notropis buchanani, N. proserpinus, Catostomus commersoni, Ictalurus bubulus, I. furcatus, I. niger, Cycleptus elongatus, Lepomis gulosus, L. megalotis, Etheostoma lepidum, Percina macrolepida*, and *Aplodinotus grunniens*. One species was introduced (*Menidia berrylina*) and another (*Notropis amabilis*) was extinct (Smith and Miller, 1986).

WATER CHEMISTRY AND BIOLOGY IN SELECTED STUDY AREAS IN TEXAS

The most detailed studies have been made in the Pecos River between Orla and Shumla. The elevation near Orla is about 832 m above mean sea level and decreases downstream until it is about 345 m above sea level near its mouth (Davis, 1980).

Invertebrates near Orla were chironomid pupae, nematodes, colonial coelenterates, oligochaetes, and ostracods. At Orla the flow is derived primarily from seepage and seasonal releases from naturally saline Red Bluff Reservoir and highly saline inflow from Screwbean Draw. This is the only flowing tributary above Sheffield.

Water quality was poor at Girvin, due to lack of fresh inflow. Therefore, the dissolved solids concentrations were high. Seepage of groundwater was saline. Irrigation return flow contributed to the Pecos in this reach. It was also possible that there were inflows of brine from oil-field activities.

The Pecos River was mildly productive (mesotrophic), as indicated by concentrations of chlorophyll *a*, total organic carbon, nitrate-nitrogen, and total phosphorus. The average total organic carbon concentrations were higher at Girvin, probably due mainly to the periodic decay of large amounts of aquatic vegetation and at Orla due mainly to excessive introduction of terrestrial plant material (Davis, 1980). A major source of nitrate-nitrogen at Orla and Girvin is probably decaying organic material, as indicated by winter peaks during maximum introduction of salt cedar leaves and concurrent elevation in amounts of ammonia nitrogen (Davis, 1980).

Based on chlorophyll *a*, total organic carbon, and the standing crop of aquatic vegetation and macroinvertebrates, the ranking order of the Texas stations in decreasing order of trophic condition was: Girvin > Orla and Sheffield > Shumla. Phosphate seemed to be the main factor limiting primary productivity in the river. This is evidenced by the fact that the density of aquatic vegetation among the stations varied directly with total phosphorus concentration and independent of nitrate-nitrogen concentration (Davis, 1980).

Turbidity was highest at Orla, probably due to the predominance of clay sediments, brown staining of water probably imparted by decaying salt cedar leaves, and moderate densities of phytoplankton. At Girvin the turbidity was lower. This was attributed to decreased velocity and increased salinity, which facilitated precipitation of suspended solids.

At Sheffield, turbidity was higher, due to lessening of salinity and increased stirring of the bottom sediments by greater velocity. At Shumla, turbidity was low, due to the limestone substrate (Davis, 1980).

Aquatic Life
The species of aquatic life occurring at Orla and Girvin had to withstand extreme salinity changes. Adaptable species often occurred in large numbers, and these must withstand extreme salinity changes. Mean diversity at Orla was 1.95 and at Girvin 1.56, which are considerably below the general range associated with clean-water streams (2.6 to 4.0). However, they are higher than the values associated with severe pollution (<1.0) (Wilhm, 1970).

At Sheffield, the number of taxa and the diversity was higher than at Orla and Girvin, indicating improved water quality. At Sheffield there were a few species usually associated with fresh water, such as *Choroterpes mexicanus* and *Stenelmis* sp. The favorable conditions at Sheffield resulted in a lower dominance pattern and lower redundancy (Davis, 1980).

Greatly improved water at Shumla resulted in a high mean diversity, a large number of taxa (50), and the presence of many typically freshwater forms, such as Ephemeroptera (eight species), dryopoid beetles (six species), and Trichoptera (nine species). The redundancy was relatively low.

Mean diversity at Shumla (3.14) was within the observed range for clean water streams, while the diversity index at Sheffield (2.5) approximated the minimum of this criterion (Davis, 1980).

"At Orla, Girvin, and Sheffield the seasonal diversity generally varied inversely with the macrobenthic standing crop, the number of taxa, and redundancy. This indicates that seasonal changes in diversity at these stations were due more to one or more taxa establishing or losing dominance than to changes in the size of the population of many species. Conversely, seasonal changes in diversity at Shumla were induced to a greater degree by changes in the number of taxa than by changes in dominance, as $\bar{d}$ varied directly with the number of taxa, and $\bar{d}$ and $\bar{r}$ varied more independently. Seasonal diversity at all stations had no direct correlation to seasonal physiochemical changes, which varied irregularly" (Davis, 1980).

Diversity at Orla varied directly with the duration of constant-volume releases from Red Bluff Reservoir prior to sampling. At Orla the diversity was usually low during irrigation seasons, when reservoir releases were irregular. This was probably due to sudden salinity fluctuations and scouring of the substrate. These conditions apparently favor a few osmotically adaptable taxa that became dominant (Davis, 1980).

"Winter peaks in diversity were usually preceded by long periods of nonreleases and stable low-flow conditions, even though salinity was high during these periods, which suggests that physiochemical fluctuations exerted more influence on diversity at Orla than did absolute chemical concentrations. Diversities at Girvin were similar in the two winter samples and in the two autumn samples, but varied widely in the spring and summer samples. Diversity seemed to be determined largely by the size of the standing crop and conditions of aquatic vegetation. Plant growth was periodically dense, but at other times it was quite sparse or absent" (Davis, 1980).

Moderate aquatic plant density in the autumn supported few taxa and individuals, and $\bar{d}$ and $\bar{r}$ were moderate. Vegetation density peaked in winter, which resulted in a large macrobenthic standing crop and higher numbers of taxa. Winter diversity was low and redundancy was high, due to the dominance of herbivores, mainly *Tanytarsus*. Vegetation was sparser during the spring, due to heavy exploitation by herbivores and nutrient depletion, the remaining vegetation being in a state of decomposition.

Diversity increased and redundancy decreased in the spring, due to the increased number of detritivores, mainly oligochaetes, and decreasing number of herbivores, mainly *Tanytarsus*. In the summer nearly all the vegetation was reduced to organic detritus. Diversity was variable and depended on the amount of food remaining. Macrobenthic standing crops were meager in July 1976 (23 m^{-2}). This was due to depletion of organic food material. In July 1977 much organic debris remained, which supported a high diversity of macrobenthos (Davis, 1980).

The predominance of decaying vegetation, which resulted in low diversity and high redundancy, promoted such feeders as *Berosus* adults (319 m^{-2}) and detrital omnivores such as Ostracoda. Temporal variations in diversity at Sheffield and Shumla were erratic, with the greatest year-to-year variation in the autumn at Sheffield and in the summer at Shumla. Diversities exceeded 3.0 twice at Sheffield and once in each season (four times) at Shumla (Davis, 1980).

The size of the macroinvertebrate standing crop seemed to be determined largely by the redundancy of aquatic vegetation. This was true at all the stations. At Orla, the year-round moderately dense aquatic vegetation supports a moderate macrobenthic standing crop which varies little in size seasonally. There were periodic dense growths of plants at Girvin. These supported large macrobenthic standing crops, which fluctuated widely in numbers (Davis, 1980).

At Orla, seasonal succession of taxa was low, as the dominant taxa exhibited a high frequency of occurrence, with 12 of the 17 taxa occurring in over half the samples. The herbivores *Ithytrichia* and *Tanytarsus*; omnivores *Cheumatopsyche*, *Dicrotendipes neomodestus*, and *Pseudochironomus*; and oligochaetes, which are organic sediment feeders, were dominant and comprised 87% of the macrobenthos.

Salt-tolerant herbivores (*Berosus* larvae, *Microtendipes*, *Tanytarsus*, *Hyalella azteca*) were present in high numbers at Girvin. Omnivorous scavengers (Ostracoda, Oligochaeta, and *Berosus* adults) were dominant during the decay of aquatic vegetation. These taxa formed 98% of the macrobenthic population. Seasonal succession of taxa was high at Girvin, as only 7 of the 21 taxa (33%) occurred in half the samples. At Sheffield 13 of the 24 taxa (52%) occurred in half the samples. Six taxa (*Pseudochironomus*, *Cricotopus bicinctus*, *Cheumatopsyche*, *Ithytrichia*, *Choroterpes mexicanus*, and Oligochaeta) comprised 87% of the macrobenthos. More carnivorous taxa were present at Sheffield than upstream: *Argia*, *Gomphus*, *Cordyalus cornutus*, *Telopelopia okoboji*, and *Tabanus*.

At Shumla, seasonal succession was high, as only 17 of the 50 taxa (34%) occurred in half the samples and the number of taxa exhibited extreme seasonal variation. This high succession was a function of the variety of life cycles represented and the resulting high rate of appearance/disappearance of species. Fifty-nine percent of the macrobenthos population was composed of *Thraulodes*, *Tricorythodes*, *Cheumatopsyche*, *Traverella*, *Cryphocricos*, and *Hydroptila*. The remaining 41% included 44 taxa. The presence of many species represented by small numbers of individuals reflects the healthy state of the community.

Few species were unique to Orla or Girvin. Several species at the upper stations were widely distributed in the river except where they were excluded by other ecological factors: high competition for limited food, high predation, and so on. This was not due to high fluctuations of salinity. Conversely, many species were restricted to lower reaches; their upstream distribution was probably limited by salinity. Salinity limits many species. For example, almost all the Ephemeroptera were freshwater species, and only one, *Choroterpes mexicanus*, was collected in these waters. Several chironomids have been considered oligohalobous (freshwater forms occurring in salinities of less than 500 mg L^{-1}). However, in the Pecos River they are present in much higher salinities, and this establishes them as euryhalinous. This group includes *Chironomus attenuatus*, *Cricotopus bicinctus*, *Dicrotendipes neomodestus*, *Goeldichironomus holoprosinus*, and *Polypedilum digitifer*.

Several species that have low tolerance to salt were represented in the Pecos River fauna by species that were either mesohalobus (occurring in salinities of 500 to 30,000 mg L^{-1}) or euryhalinous. These Pecos forms may include differentiated undescribed species that have evolved to exist under saline conditions. This group includes *Ablabesmyia*, *Cladotanytarsus*, *Microtendipes*, *Pseudochironomus*, and *Tanytarsus*. Many of the species at Shumla were considered freshwater forms, but since at Shumla salinity frequently exceeds 500 mg L^{-1}, they cannot be considered truly oligohalobous species. *Paranais litoralis*, usually considered a saltwater form, occurred in the Pecos River at Girvin. This is only the third time it has been recorded as being found inland in the world.

"A rare Oligochaete *Monopylephorus* sp. was collected at Orla, Girvin, and Sheffield, being known previously from swamps in southeastern Louisiana. *Monopylephorus* sp. occurs among the roots of aquatic plants, and its presence appears to be associated with the coastal macrophyte found extensively in the upper Pecos, widgeon grass (*Ruppia maritima*)" (Davis, 1980).

"Two ostracods were found at Girvin: *Cyprideis* sp., a nonswimming genus characteristic of brackish water, and *Limnocythere* sp., a minute mud-crawling form morphologically similar to a species found on Long Island. The naucorid hemipterans *Cryphocricos* and *Limnocoris* were common in the lower river, each having been reported previously from single locations in the United States. The rare trichopteran *Ithytrichia* was common in the upper reaches, and specimens contributed to the National Museum of Natural History were the first larvae to be deposited there" (Davis, 1980).

"The colonial coelenterate, *Corydalus lacustris*, was collected in Orla and Sheffield, and these are among the first Texas records. The Asiatic clam, *Corbicula manilensis*, was collected at Shumla. This was the first record of this widely distributed exotic in the Pecos River. This pelecypod is common below Shumla in the Amistad Reservoir and the adjacent Rio Grande. The freshwater sponge, *Spongilla* sp., was collected at Shumla and supported the parasitic neuropteran, *Climacia areolaris*" (Davis, 1980).

Structure of Ecosystems

The structure of the ecosystems in the Pecos River consists of four stages of nutrient and energy transfer (detritus, algae; detritivores, herbivores; omnivores; carnivores) as set forth by Eugene P. and Howard T. Odum.

The structure of the ecosystem in the Pecos River is derived largely from the work of Rhodes and Hubbs (1992) and from Davis (1980). (Table 4.7.)

TABLE 4.6.   *Families Represented by Species: Vegetation of Watershed.*

DIVISION
MAGNOLIOPHYTA
   Ruppia maritima
CLASS
DICOTYLEDONEAE
  ORDER LAURALES
    Family Lauraceae
  ORDER NYMPHAEALES
    Family Ceratophyllaceae
    Family Nelumbonaceae
    Family Nymphaeaceae
  ORDER
  RANUNCULALES
    Family Ranunculaceae
  ORDER
  HAMAMELIDALES
    Family Plantaceae
  ORDER URTICALES
    Family Ulmaceae
    Family Urticaceae
  ORDER
  CARYOPHYLLALES
    Family Aizoaceae
    Family Amarnathaceae
    Family Caryophaceae
    Family Phytolaccaceae
  ORDER POLYGONALES
    Family Polygonaceae
  ORDER THEALES
    Family Clusiaceae
    (=Guttiferaceae)
    Family Elatinaceae
  ORDER MALVALES
    Family Malvaceae
  ORDER NEPENTHALES
    Family Droseraceae
  ORDER SALICALES
    Family Salicaceae
  ORDER CAPPARALES
    Family Brassicaceae
  ORDER BATALES
    Family Bataceae
  ORDER ERICALES
    Family Ericaceae
  ORDER PRIMULALES
    Family Primulaceae

ORDER ROSALES
  Family Rosaceae
ORDER FABACEAE
  Family Caesalpiniaceae
  Family Fabaceae
  Family Mimosaceae
ORDER
PODOSTEMANIALES
  Family Podostemaceae
ORDER HALORAGALES
  Family Haloragaceae
ORDER MYRTALES
  Family Lythraceae
  Family Melastomataceae
  Family Onagraceae
ORDER CORNALES
  Family Cornaceae
ORDER EUPHORBIALES
  Family Euphorbiaceae
ORDER SAPINDALES
  Family Aceraceae
ORDER GERANIALES
  Family Asclepiadaceae
  Family Balsaminaceae
ORDER APIALES
  Family Apiaceae
  (=Umbeliferaceae)
ORDER GENTIANALES
  Family Asclepiadaceae
  Family Gentianaceae
  Family Loganiaceae
ORDER SOLONALES
  Family Convolvulaceae
  Family Solanaceae
ORDER LAMIALES
  Family Boraginaceae
  Family Verbenaceae
ORDER
CALLITRICHALES
  Family Callitrichaceae
ORDER
SCROPHULARIALES
  Family Acanthaceae
  Family Lentibulariaceae
  Family Scrophulariaceae

ORDER
CAMPANULALES
  Family Campanulaceae
ORDER RUBIALES
  Family Rubiaceae
ORDER ASTERALES
  Family Asteraceae
CLASS
MONOCOTYLEDONEAE
  ORDER ALISMALES
    Family Alismaceae
    Family Butomaceae
  ORDER
  HYDROCHARITALES
    Family
    Hydrocharitaceae
  ORDER NAJADALES
    Family Najadaceae
    Family
    Potamogetonaceae
    Potomogeton sp.
    Family Zannichelliaceae
  ORDER ARALES
    Family Araceae
    Family Lemnaceae
  ORDER
  COMMELINALES
    Family Xyriceae
  ORDER ERIOCAULES
    Family Eriocaulaceae
  ORDER JUNCALES
    Family Juncaceae
  ORDER CYPERALES
    Family Cyperaceae
    Family Poaceae
    Phragmites sp.
  ORDER TYPHALES
    Family Sparganiaceae
    Family Typhaceae
  ORDER ILIALES
    Family Iliaceae
    Family Iridaceae
    Family Pontederiaceae
  ORDER ORCHIDALES
    Family Orchidaceae

TABLE 4.7.   *Species List: Aquatic Life in the Middle and Lower Pecos River*

Taxon[a]	Substrate[b]			Site[c]	Source[d]
	V,D	M,S	G,R		
SUPERKINGDOM PROKARYOTAE					
KINGDOM MONERA					
Division Cyanophycota					
Class Cyanophyceae					
Order Chroococcales					
Family Chroococcaceae					
Family Entophysalidaceae					
Order Chamaesiphonales					
Family Chamaesiphonaceae					
Family Clastidiaceae					
Order Stignematales					
Family Stigonemataceae					
Order Nostocales					
Family Nostocaceae					
Family Oscillatoriaceae					
Family Rivulariaceae					
Family Scytonemataceae					
Family Stigonemaceae					
SUPERKINGDOM EUKARYOTAE					
KINGDOM PLANTAE					
Division Rhodophycota					
Class Rhodophyceae					
Order Bangiales					
Family Bangiaceae					
Order Compsopongonales					
Family Compospogonaceae					
Order Acrochaetinales					
Family Acrochaetiaceae					
Family Audouinellaceae					
Order Nemalionales					
Family Batrachospermaceae					
Family Lemaneaceae					
Division Chromophycota					
Class Xanthophyceae					
Order Vaucheriales					
Family Vaucheriaceae					
* *Vaucheria* sp.	X	X			b
Class Bacillariophyceae					
Order Eupodiscales					
Family Coscinodiscaceae					
Family Thalassiosiraceae					
Order Biddulphiales					
Family Anaulaceae					
Family Biddulphiaceae					
Order Fragilariales					

TABLE 4.7. (*Continued*)

Taxon[a]	Substrate[b]			Site[c]	Source[d]
	V,D	M,S	G,R		
Family Fragilariaceae					
Synedra sp.				S	b
Order Eunotiales					
Family Eunotiaceae					
Order Achnanthales					
Family Achnanthaceae					
Order Naviculales					
Family Cymbellaceae					
Cymbella sp.				L	b
Family Entomoneidaceae					
Family Gomphonemaceae					
Family Naviculaceae					
Order Epithemiales					
Family Epithemiaceae					
Order Bacillariales					
Family Nitzschiaceae					
Order Surirellales					
Family Surirellaceae					
Class Dinophyceae					
Order Gymnodiniales					
Family Gymnodiniaceae					
Order Peridinales					
Family Peridiniaceae					
Family Pyrophacaceae					
Order Phytodinales					
Family Phytodiniaceae					
Division Haptophyta					
Class Prymnesiaphyceae					
Order Prymnesiales					
Family Prymnesiaceae					
Prymnesium parva					
Division Chlorophycota					
Class Euchlorophyceae					
Order Tetrasporales					
Family Chlorangiaceae					
Family Tetrasporaceae					
Order Volvocales					
Family Chlamydomonadaceae					
Family Haematoccaleae					
Family Phacotaceae					
Family Spondylomoraceae					
Family Volvocaceae					
Order Chlorococcales					
Family Chloroccaceae					
Family Coccomyxaceae					

(*continued*)

TABLE 4.7.   (*Continued*)

Taxon[a]	Substrate[b]			Site[c]	Source[d]
	V,D	M,S	G,R		
Family Hydrodictyaceae					
Family Oocystaceae					
Family Palmellaceae					
Family Scenedesmaceae					
Class Ulotricophyceae					
Order Ulotrichales					
Family Cylindrocapsaceae					
Family Microsporaceae					
Family Trentepohliaceae					
Family Ulothrichaceae					
Ulothrix sp.	X	X		S	b
Order Ulvales					
Family Ulvaceae					
Order Prasiolales					
Family Prasiolaceae					
Order Chaetophorales					
Family Chaetophoraceae					
Family Schizomeridaceae					
Order Oedogoniales					
Family Oedogoniaceae					
Order Siphonocladales					
Family Cladophoraceae					
* *Cladophora* sp.	X		X	S	b
Order Siphonales					
Family Phyllosiphonaceae					
Class Zygophyceae					
Order Zygnematales					
Family Desmidiaceae					
Family Mesotaeniaceae					
Family Zygnemataceae					
* *Zygnema* sp.	X	X		L	b
Class Charophyceae					
Order Charales					
Family Characeae					
* *Chara* sp.	X	X		L	b
Class Monocotyledoneae					
Order Najadales					
Tribe Potamogetonaceae					
Potamogeten				L	b
KINGDOM ANIMALIA					
Phylum Porifera					
Class Demospongiae					
Order Haplosclerida					
Family Spongillidae					
unident. sp.				L	b
Phylum Cnideria					

TABLE 4.7.   (*continued*)

Taxon[a]	Substrate[b]			Site[c]	Source[d]
	V,D	M,S	G,R		
Class Hydrozoa					
Order Hydroida					
Family Clavidae					
*o *Cordylophora lacustris*	X	X		S,L	b
Phylum Platyhelminthes					
Class Turbellaria					
Order Tricladida					
Family Planariidae					
*do *Dugesia tigrina*	X	X		L	b
Phylum Nemata					
unident. spp.				S,L	b
Phylum Mollusca					
Class Gastropoda					
Order Mesogastropoda					
Family Hydrobiidae					
Family Pleuroceridae					
Family Valvatidae					
Family Viviparidae					
Order Basommatophora					
Family Ancylidae					
Family Lymnaeidae					
Family Physidae					
* *Physa virgata*	X	X		S	b
Family Planorbidae					
Class Bivalvia					
Order Unionoida					
Family Unionidae		X			
Order Veneroida					
Family Corbiculidae					
*o *Corbicula manilensis (= flumenea?)*		X		L	b
Family Sphaeriidae					
*o *Sphaerium striatinum*		X		L	b
Phylum Annelida					
Class Oligochaeta					
Order Lumbriculida					
Family Lumbriculidae					
unident. spp.	X	X		L	b
Order Haplotaxida					
Family Naididae					
* *Paranais litoralis*		X		L	b
Family Tubificidae					
*do *Limnodrilus hoffmeisteri*	X	X		S	b
*do *L. udekemianus*	X	X		S	b
*do *Monopylephorus* sp.	X	X		S,L	b
Phylum Arthropoda					

(*continued*)

TABLE 4.7.   (*Continued*)

Taxon[a]	Substrate[b] V,D	M,S	G,R	Site[c]	Source[d]
Class Arachnida					
Order Acariformes					
Family Arrenuridae					
* *Arrenurus* sp.	X	X		S	b
Class Crustacea					
Subclass Ostracoda					
Order Podocopina					
Family Cytheridae					
* *Cyprideis* sp.	X	X		S	b
* *Limnocythere* sp.	X	X		S	b
Subclass Malacostraca					
Order Amphipoda					
Family Gammaridae					
o unident. sp.	X	X	X	M	k
Family Talitridae					
*o *Hyalella azteca*		X	X	S	b
Class Insecta					
Order Odonata					
Suborder Zygoptera					
Family Calopterygidae					
*c *Hetaerina* sp.	X		X	M	k
Family Coenagrionidae					
*c *Argia translata*	X		X	S	b
*c *Argia* sp.	X		X	S,L	b
*c *Enallagma* sp.	X		X	S	b
*c *Ischneura* sp.	X		X	M	k
Suborder Anisoptera					
Family Gomphidae					
*c *Erpetogomphus* sp.		X	X	M,S,L	b,k
*c *Gomphus externus*		X		S	b
*c *Gomphus* sp.		X		M	k
*c *Ophiogomphus* sp.			X	M	k
Family Libelludidae					
c *Brechmorhaga mendax*				L	b
unident. sp.			X	S	b
Order Ephemeroptera					
Suborder Schistonota					
Family Baetidae					
*do *Baetis* sp.	X	X	X	M,L	b,k
* *Baetodes* sp.			X	L	b
*o *Callibaetis* sp.	X		X	M	k
Family Heptageniidae					
*o *Heptagenia* sp.		X	X	M	k
Family Leptophlebiidae					
* *Choroterpes mexicanus*			X	L,S	b

TABLE 4.7. (*Continued*)

Taxon[a]	Substrate[b]			Site[c]	Source[d]
	V,D	M,S	G,R		
Family Leptophlebiidae (*cont.*)					
* *Thraulodes* sp.			X	M,L	b,k
* *Traverella* sp.			X	L	b
Family Oligoneuriidae					
*co *Isonychia* sp.	X		X	L	b
Suborder Pannota					
Family Caenidae					
*o *Caenis* sp.	X	X		L	b
Family Tricorythidae					
*o *Tricorythodes* sp.	X	X	X	M,L	b,k
Order Plecoptera					
Family Perlidae					
*c *Acroneuria abnormis*		X	X	M	k
Order Hemiptera					
Family Corixidae					
unident. spp.				S	b
Family Naucoridae					
*c *Ambrysus mormona*	X	X	X	M	k
*c *Ambrysus* sp.	X	X	X	L	b
*c *Cryphocricos* sp.		X	X	S,L	b
*c *Limnocoris* sp.		X	X	L	b
Order Neuroptera					
Family Sisyridae					
Climacea areolaris				L	b
Order Megaloptera					
Family Corydalidae					
*c *Corydalus cornutus*	X	X	X	S,L	b
Order Coleoptera					
Suborder Adephaga					
Family Haliplidae					
*h *Peltodytes* sp.	X	X		M	k
Suborder Polyphaga					
Family Dryopidae					
*o *Helichus* sp.			X	M	k
Family Elmidae					
o *Elsianus texanus*				L	b
*o *Hexacylloepus ferrugineus*			X	L	b
*d *Microcylloepus pusillus*			X	L	b
*o *Stenelmis bicarinata*			X	L	b
*o *Stenelmis* sp.		X	X	S	b
Family Hydrophilidae					
*h *Berosus infuscatus*		X		S	b
*h *B. subsignatus*		X		S	b
*h *Berosus* spp.		X		S	b
Family Limnichidae					
*o *Lutrochus luteus*			X	L	b

(*continued*)

TABLE 4.7. (*Continued*)

Taxon[a]	Substrate[b] V,D	M,S	G,R	Site[c]	Source[d]
Family Psephenidae					
*o *Psephenus texanus*			X	L	b
Order Trichoptera					
Family Glossosomatidae					
*o *Protoptila* sp.			X	L	b
Family Hydropsychidae					
*o *Cheumatopsyche* sp.	X		X	S,L	b
*o *Hydropsyche* sp.	X		X	M,L	b,k
*o *Smicridea* sp.		X	X	L	b
Family Hydroptilidae					
*h *Hydroptila* sp.		X	X	M,S,L	b,k
*o *Ithytrichia* sp.			X	S	b
o *Mayatrichia* sp.				M	k
h *Stactobiella* sp.				M	k
unident. sp.			X	L	b
Family Leptoceridae					
*ho *Nectopsyche gracilis*	X	X	X	M	k
*o *Oecetis avara*		X	X	L	b
Family Philopotamidae					
*o *Chimarra* spp.			X	L	b
Family Polycentropodinae					
*c *Cernotina?* sp.			X	L	b
Order Lepidoptera					
Family Pyralidae					
*h *Cataclysta* sp.	X	X		S,L	b
Order Diptera					
Suborder Nematocera					
Family Ceratopogonidae					
*o *Dasyhelea* sp.	X	X	X	L	b
*c *Palpomyia tibialis*	X	X	X	S	b
unident. sp.				M,L	b,k
Family Chironomidae					
*o *Ablabesmyia* sp.		X	X	M,S	b,k
*o *Calopsectra* sp.			X	M	k
ho *Chironomus attenuatus*				S	b
*ho *Chironomus* sp.	X	X		M,S	b,k
*o *Cladotanytarsus* sp.	X	X	X	S	b
*o *Cricotopus bicinctus*	X	X		S,L	b
*o *Cricotopus* spp.	X	X		L	b
*o *Cryptochironomus* sp.		X	X	M	k
*o *Dicrotendipes neomodestus*		X		S	b
*o *Goeldichironomus holoprasinus*		X		S	b
*o *Microtendipes* sp.		X	X	S	b
*o *Nilotanypus* sp.			X	L	b
*o *Polypedilum digitifer*	X	X		S	b
*o *Polypedilum* spp.	X	X		M,S,L	b,k

TABLE 4.7. (*Continued*)

Taxon[a]	Substrate[b]			Site[c]	Source[d]
	V,D	M,S	G,R		
Family Chironomidae (*cont.*)					
*c *Procladius* sp.		X		M	k
*c *Psectrotanypus* sp.			X	L	b
*o *Pseudochironomus* sp.		X	X	S,L	b
*o *Rheotanytarsus* sp.			X	L	b
Tanytarsus spp.				S,L	b
*c *Telopelopia okoboji*			X	S	b
unident. spp.				S,L	b
Family Simuliidae					
*o *Simulium venustum*			X	L	b
*do *Simulium* sp.			X	M	k
Family Tipulidae					
*c *Limnophila* sp.		X	X	M	k
Suborder Brachycera					
Family Empididae					
*co *Hemerodromia* sp.			X	M	k
unident. sp.				L	b
Family Muscidae					
*c unident. Anthomyiidae		X	X	S	b
Family Stratiomyidae					
Odontomyia sp.				M	k
Family Tabanidae					
*c *Crysops* sp.		X	X	M	k
*c *Tabanus* sp.		X	X	L	b
Phylum Chordata					
Subphylum Vertebrata					
Class Osteichthyes					
Order Lepisosteiformes					
Family Lepisosteidae					
*c *Lepisosteus osseus*	X	X	X	M	f
Order Clupeiformes					
Family Clupeidae					
*o *Dorosoma cepedianum*	X	X	X	M,S,L	f,h
Order Cypriniformes					
Family Catostomidae					
*c *Carpiodes carpio*		X		M,L	f,h
c *Moxostoma congestum*				L	h
Family Cyprinidae					
*c *Cyprinella (= Notropis) lutrensis*		X	X	S,L	h
o C. (= *Notropis) proserpina*				S,L	h
*c C. (= *Notropis) venusta*		X		L	h
*o *Cyprinus carpio*		X		M,S,L	f,h
*h *Dionda episcopa*		X		S,L	h
*h *Hybognathus placitus*		X	X	M	f
*o *Hybopsis aestivalis*		X	X	M,L	f,h
Notropis amabilis				L	h

(*continued*)

TABLE 4.7.   (*Continued*)

Taxon[a]	Substrate[b]			Site[c]	Source[d]
	V,D	M,S	G,R		
Family Cyprinidae (*cont.*)					
N. braytoni				L	h
*o N. jemezanus		X	X	M,L	f,h
*o N. lutrensis		X	X	M	f
*o N. simus	X	X	X	M	f
*co N. stramineus		X		M	f
*co Phenacobius mirabilis		X	X	M	f
*o Pimephales promelas		X		M,L	f,h
*o P. vigilax		X		L	h
*o Rhinichthys cataractae			X	M	f
*c Semotilus atromaculatus			X	M	f
Order Characiformes					
Family Characidae					
*co Astyanax mexicanus			X	S,L	h
Order Siluriformes					
Family Ictaluridae					
*o Ictalurus lupus		X		M,L	k
*o I. punctatus		X	X	M,L	f,h
*c Pylodictis olivaris		X		M,L	f,h
Order Percopsiformes					
Family Amblyopsidae					
Family Aphredoderidae					
Family Percopsidae					
Order Cyprinodontiformes					
Family Cyprinodontidae					
Cyprinodon atomus					
C. diabolis					
C. maculatus					
C. nevadensis					
*o Cyprinodon pecosensis			X	M, S, L	f
C. rubrofluviatus					
C. variegetus					
*o Cyprinodon hybrids			X	S, L	h
*o Fundulus grandis		X		L	h
o F. zebrinus				M,S	f,h
* Lucania parva	X			S,L	h
Family Poeciliidae					
*o Gambusia affinis	X	X		M,S,L	f,h
Order Atheriniformes					
Family Atherinidae					
*c Menidia beryllina		X	X	S,L	h
Order Perciformes					

TABLE 4.7. (*Continued*)

Taxon[a]	Substrate[b]			Site[c]	Source[d]
	V,D	M,S	G,R		
Family Centrarchidae					
*c *Lepomis cyanellus*		X	X	M,S,L	f,h
*c *L. macrochirus*			X	S	h
*c *L. megalotis*	X	X		S,L	h
*c *Micropterus salmoides*	X		X	L	h
Family Cichlidae					
Cichlasoma cyanoguttatum				L	h
Family Percidae					
c *Etheostoma grahami*				L	h
Family Percithyidae					
Family Sciaenidae					

*Species found in the ecosystem described in the section "Structure of the Ecosystem."

[a]c, Carnivore; co, carnivore-omnivore; d, detritivore; do, detritivore-omnivore (devours organisms and all types of detritus; h, herbivore; ho, herbivore-omnivore; o, omnivore.

[b]V, vegetation; D, detritus; M, mud; S, sand; G, gravel; R, rock.

[c]M, middle Pecos River (between Santa Rosa, NM and Malaga, NM); S, saline reach (between Malaga, NM and Sheffield, TX); L, lower Pecos River (between Sheffield, TX and confluence with the Rio Grande at Amistad Reservoir).

[d]b, Davis (1980) (Pecos River); f, Hatch et al. (1985) (middle Pecos); h, Rhodes and Hubbs (1992) (saline and lower Pecos); k, Kelsch and Hendricks (1990).

Associated with Vegetation and Detritus. *Vaucheria* was found associated with the substrate and the bases of vegetation, such as stems and exposed roots. *Cladophora* was also found in this habitat associated with the vegetation: stems, sediments, and leaves. *Chara* was found growing in and among the vegetation.

Of the invertebrates, *Dugesia tigrina* was found crawling over the sediments, as was the snail, *Physa virgata*. Living attached to the vegetation was the hydroid *Cordylophora lacustris*, which seems to be an omnivore. Various worms were found in and among the vegetation. They were an unidentified species of the earthworm family Lumbriculidae. Of the Tubificid family, there were the detritivore-omnivores *Limnodrilus hoffmeisteri*, *L. udekemianus*, and an unidentified species of the genus *Monopylephorus*. The arachnid *Arrenurus* sp. was also found on the vegetation, as were the crustaceans *Cyprideis* sp. and *Limnocythere* sp. An undetermined species of amphipod, an omnivore, was found in and among the vegetation.

Dragonflies found in this habitat were the carnivores *Hetaerina* sp., *Argia translata*, *Argia* sp., *Enallagma* sp., and *Ischneura* sp. Several mayflies were found in and among the vegetation: *Baetis* sp., which is a detritivore-omnivore, and *Callibaetis* sp., an omnivore. Other mayflies found were the carnivore-omnivore *Isonychia* sp., and *Caenis* sp. and *Tricorythodes*, omnivores. All of these were living on or among the vegetation. Two hemipterans were also found in this habitat: the carnivorous *Ambrysus mormona* and an unidentified species of the same genus. The megalopteran *Corydalus cornutus*, a carnivore, was found on the vegetation. Associated with the vegetation was the beetle *Peltodytes*, a herbivore. Living attached to the vegetation where the water was flowing rather swiftly were the Hydropsychidae; *Cheumatopsyche*, an omnivore; and *Hydropsyche*, an omnivore. A herbivore-omnivore in this habitat was *Nectopsyche gracilis*. One lepidopteran was found in this habitat, the herbivore, *Cataclysta* sp.

Several dipterans were common living on the vegetation or on the surface of the substrate in the debris associated with the vegetation. They were the omnivorous *Dasyhelea* sp. and the carnivore *Palpomyia tibialis*. The chironomids found here were the herbivore-omnivore *Chironomus* sp., the omnivores *Cladotanytarsus* sp., *Cricotopus bicinctus*, and unidentified species of the same genus, *Polypedilum digitifer* and *Polypedilum* spp.

Several fish were found associated with the vegetation. They were the carnivorous *Lepisosteus osseus* and the omnivorous *Dorosoma cepedianum*. Other species associated with the vegetation were *Cyprinus carpio* and *Notropis simus*, both omnivores. In and among the vegetation was *Lucania parva*. Swimming in and among the vegetation was *Gambusia affinis*, an omnivore. The carnivores *Lepomis megalotis* and *Micropterus salmoides* were found in this habitat.

Associated with Mud and Sand. A common habitat in the Pecos River is mud and sand found in pools and slack waters and sometimes along the edges of the stream. Species living in these habitats are varied. Among the algae, *Vaucheria* sp. is often found attached to mud or sand and sometimes attached to rocks in slowly flowing water. In this habitat an unidentified species of the genus *Zygnema* might be found. Also growing in pools or slack waters one sometimes finds unidentified species of *Chara*. Hydra may be found attached to debris. The omnivorous *Cordylophora lacustris* was found here. Crawling over the debris in pools or slack water is the detritivore-omnivore, *Dugesia tigrina*. Other species found in this habitat were the snail, *Physa virgata*, the omnivorous *Corbicula manilensis*, and in the substrate particularly among gravel may be found *Sphaerium striatinum*, an omnivore. Various worms were found in this habitat. Unidentified species of the family Lumbriculidae were found. *Paranais litoralis* was also found along with the detritivore-omnivores *Limnodrilus hoffmeisteri*, *L. udekemianus*, and *Monopylephorus* sp. The arachnid *Arrenurus* sp. was found in among the debris. Also present were the ostracods *Cyprideis* sp. and *Limnocythere* sp. These taxons are probably omnivorous although a true definition of their food preferences was not found. Other species found in the slack water, particularly in pools, were the amphipod belonging to the family Gammaridae and *Hyalella azteca*, an omnivore.

Quite a few insects were found in and among the debris. Particularly in pools, but sometimes associated with larger materials such as logs or stems of plants, were the Anisoptera, *Erpetogomphus* sp., a carnivore; *Gomphus externus*, a carnivore; and an unidentified species of *Gomphus*, also probably a carnivore. Several mayflies were also found in this habitat. They were an unidentified species of the genus *Baetis*, probably a detritivore-omnivore, and unidentified species of the genus *Heptagenia*, an omnivore. Other Ephemeroptera found in this habitat, particularly in pools, were the omnivorous *Caenis* sp. and *Tricorythodes* sp. The stonefly, *Acroneuria abnormis*, which is a carnivore, was found in pools where the current was fairly rapid. Certain Hemiptera were also found in this habitat: the carnivorous *Ambrysus mormona*, *Ambrysus* sp., *Cryphocricos* sp., and *Limnocoris* sp. Another carnivore found was the megalopteran *Corydalus cornutus*. An unidentified species of *Peltodytes*, a herbivore, was found. Several beetles were found in among the debris in this habitat. They were the omnivorous *Stenelmis* sp., the herbivores *Berosus infuscatus* and *B. subsignatus*, and unidentified species of the same genus, also probably herbivorous. Another species belonging to the Coleoptera was *Lutrochus luteus*, an omnivore.

The trichopterans *Smicridea* sp., an omnivore, and the herbivore *Hydroptila* sp. were found. Other Trichoptera found in this habitat belong to the family Leptoceridae. They were the herbivore-omnivore *Nectopsyche gracilis* and the omnivore *Oecetis avara*. Another species in this habitat was a species of the order Lepidoptera: *Cataclysta* sp., a herbivore. Several Diptera were found in areas where the current was relatively slow. They are common among the detritus. Here were found the omnivore *Dasyhelea* sp. and the carnivore *Palpomyia tibialis*. Chironomids in this habitat were the omnivore *Ablabesmyia* sp.; *Chironomus* (an unknown species), probably a herbivore-omnivore; and the omnivorous *Cladotanytarsus* sp., *Cricotopus bicinctus*, *Cricotopus* spp., *Cryptochironomus* sp., *Dicrotendipes neomodestus*, *Goeldichironomus holoprasinus*, *Microtendipes* sp., *Polypedilum digitifer*, and unidentified species of the same genus. Also found were an unidentified species of the genus *Procladius*, a carnivore, and the omnivorous *Pseudochironomus* sp. The tipulid larvae *Limnophila* sp., a carnivore, was found among the detritus. Undetermined species of the family Anthomyiidae were found. A few tabanids were in this habitat where the current was relatively slow, particularly in pools. Here one found the carnivores *Chrysops* sp. and *Tabanus* sp.

The pools and slackwater were favorable habitats for a number of species of fish. They were *Lepisosteus osseus*, a carnivore; *Dorosoma cepedianum*, an omnivore; *Carpiodes carpio*, a carnivore; and *Cyprinella lutrensis*, a carnivore. Also found in these pools and sometimes in the slack water were the carnivorous *Cyprinella venusta*. A herbivore, *Dionda episcopa*, was found here, as was *Hybognathus placitus*, also a herbivore, and the omnivore *Hybopsis aestivalis*. Other cyprinids found in the pools and slack water were the omnivores *Notropis jemezanus*, *N. lutrensis*, and *N. simus*. These slack waters, sometimes backwaters, were favorable habitats for many species of fish. Here one found the carnivore-omnivores *Notropis stramineus* and *Phenacobius mirabilis* and the omnivores *Pimephales promelas* and *P. vigilax*. Catfish were also found in these slack water habitats and in pools: *Ictalurus lupus*, an omnivore; *I. punctatus*, an omnivore; and *Pylodictis olivaris*, a carnivore. Other fish found in pools or slack waters were the omnivores *Fundulus grandis* and *Gambusia affinis*. The carnivorous *Menidia beryllina* was also found in these pools. Several centrarchids were encountered in these slack water habitats: the carnivorous *Lepomis cyanellus* and *L. megalotis*.

Lotic Habitats.   In rapidly flowing water where the substrate is rocks, pebbles, and gravel (some sand and mud may also be present), one finds *Cladophora* sp. attached to the rocks. Among the rocks where the current is not as strong were unidentified amphipods, which are probably omnivores. Also present was *Hyalella azteca*, which is an omnivore.

Of the insects, the carnivorous *Hetaerina* sp., an odonate, was present in between the rocks and where the current was not as strong. Also in these habitats, where shear force of current is not present, one finds *Argia translata* and an unidentified species of *Argia*, an unidentified species of the genus *Enallagma* sp., and *Ischneura* sp., all of which are carnivores. On the surface of the rocks and underneath the rocks were the gomphids, *Erpetogomphus* sp. and *Ophiogomphus* sp., both carnivores. An unidentified species of Libellulidae was present. Mayflies were represented by the detritivore-omnivore *Baetis* sp.; *Baetodes* sp., probably a detritivore-omnivore; and *Callibaetis* sp., an omnivore. Another mayfly present in this habitat was *Heptagenia* sp., an omnivore. Belonging to the family Leptophlebiidae were *Choroterpes mexicanus*, *Thraulodes* sp., and *Traverella* sp. Also present in this habitat was the carnivore-omnivore *Isonychia* sp. The family Tricorythidae was represented by an unidentified species of the genus *Tricorythodes*, an omnivore. The stoneflies present were *Acroneuria abnormis*, a carnivore. Several bugs were found: the carnivorous *Ambrysus mormona*, *Ambrysus* sp., *Cryphocricos* sp., and *Limnocoris* sp. The Megaloptera were represented by *Corydalus cornutus*, a carnivore. *Helichus* sp., an omnivore, was also present. The Elmidae beetles were represented by the omnivores *Hexacylloepus ferrugineus*, *Stenelmis bicarinata*, and *Stenelmis* sp. and the detritivore *Microcylloepus pusillus*. These beetles were found largely in between the rocks or in the surface substrate. Another omnivore present was *Psephenus texanus*, also found among the rocks and associated with any debris that might have been caught in this habitat. The omnivorous trichopteran, *Protoptila* sp., was also in this habitat. Several hydropsychids were found attached to the rocks in this habitat: the omnivores *Cheumatopsyche* sp., *Hydropsyche* sp., and *Smicridea* sp. The herbivore *Hydroptila* sp. was also found in among the rocks. The omnivore *Ithytrichia* sp. was also in this habitat, as were two members of the family Leptoceridae, the herbivore-omnivore *Nectopsyche gracilis* and the omnivore *Oecetis avara*. Unidentified species of the genus *Chimarra*, omnivores, were also present in and among the rocks, as was the carnivore of the family Polycentropodinae, *Cernotina?* sp. Two ceraptogonids were found where the current is relatively fast. They were the omnivore *Dasyhelea* sp. and the carnivore *Palpomyia tibialis*. Chironomids were found in the sand in this habitat: the omnivores *Ablabesmyia* sp. and *Calopsectra* sp. Several chironomids were in among the sand and vegetation associated with this habitat: the omnivores *Cladotanytarsus* sp., *Cryptochironomus* sp., *Microtendipes* sp., and *Nilotanypus* sp. Other chironomids that were omnivores found among the debris and the sand associated with the rocks in relatively fast-flowing water were *Pseudochironomus* sp. and *Rheotanytarsus* sp. *Psectrotanypus* sp. and *Telopelopia okoboji* are carnivores found here. Where water was flowing relatively fast, there were several blackflies.

They belong to at least two species, the omnivore *Simulium venustum* and an uniden-tified species of *Simulium*, a detritivore-omnivore. Also in this habitat was *Limnophila* sp., a carnivore. Other insects found in and among the rocks were the car-nivore-omnivore *Hemerodromia* and an unidentified species of the family Anthomyi-idae, which is a carnivore. Two Tabanidae were found. They were the carnivore *Chrysops* sp. and the carnivore *Tabanus* sp. They were in and among the rocks.

These areas of relatively fast-flowing water were favorable habitats for several species of fish. These fish included *Lepisosteus osseus*, a carnivore; *Dorosoma ce-pedianum*, an omnivore; *Cyprinella lutrensis*, a carnivore; *Hybognathus placitus*, a herbivore; and *Hybopsis aestivalis*, an omnivore. Cyprinids belonging to the genus *Notropis* found in this habitat included *Notropis jemezanus*, *N. lutrensis*, and *N. simus*, all omnivores. Other cyprinids were *Phenacobius mirabilis*, a carnivore-omnivore; *Rhinichthys cataractae*, an omnivore; and *Semotilus atromaculatus*, a car-nivore. Also present was the carnivore-omnivore *Astyanax mexicanus* and the omnivore *Ictalurus punctatus*. Several cyprinodonts were found in this habitat: the omnivorous *Cyprinodon pecosensis* and *Cyprinodon* hybrids. Other species found in the rheophilic conditions where rocks and gravel had accumulated were *Menidia beryllina*, a carnivore; and the centrarchids *Lepomis cyanellus*, *L. macrochirus*, and *Micropterus salmoides*, all carnivores.

SUMMARY

The Pecos River originates in the high mountains of the Sangre de Cristo range, where it is fed by snowmelt and rainfall. The river is characterized from its origin to Fort Sumner, New Mexico as a clear freshwater mountain stream. In New Mexico it begins to accumulate dissolved Permian salts. These contribute to the saline nature of the river. Added to the river at Malaga are additional salts that arise from saline springs.

Once the Pecos enters Texas, it flows into Red Bluff Reservoir. This reservoir has a capacity of about 310,000 acre-ft. Red Bluff Reservoir is used for flood control and for irrigation. The river increases in salinity as it meanders from Red Bluff Reservoir to Girvin. This increase is caused by the dissolution of Permian salts and the high evaporation rate due to the arid climate.

The river flows out of the Permian formation into the Cretaceous limestone of the Edwards–Stockton Plateau as it reaches Sheffield. Freshwater flow is increased in this area from groundwater springs and tributaries, which dilute the salts in the river.

The major tributaries in the upper river are Cow Creek, Rio Mora, Gallinas River, Tecolote Creek, Cañon Blanco, Pintada Arroyo, and Alamogordo Creek. Between Sumner Lake and Carlsbad, the main tributaries are Rio Hondo, Rio Felix, Rio Pe-nasco, and Seven Rivers. In the Carlsbad area the principal drainage includes Dark Canyon, Black River, and the Delaware River. The Pecos River is regulated by Sum-ner Lake, Two Rivers Reservoir, Lake McMillan, and Red Bluff Reservoir.

The climate of the basin becomes more arid as one progresses from north to south. The New Mexico watershed of the Pecos River is approximately 25,992 mi^2. The Texas watershed is approximately 12,000 mi^2.

The aquifers in the basin are primarily unconsolidated sand and gravel, limestone and sandstone in New Mexico. Aquifers in the Pecos basin are typically limestone or alluvium in general. They are connected to the river hydrologically. This is particularly true in New Mexico. The Texas Pecos watershed is an arid region with average rainfall that increases downstream from 10 cm at Orla to 41 cm at the mouth. Most of the geology of the basin is dominated by sedimentary deposits. Igneous rocks are most plentiful in the western basin, particularly in the mountains. Most of the water-bearing rocks are located in the plains. Water-bearing deposits of limestone, anhydrite, and halite are plentiful and commonly influence water chemistry.

The main land use of the watershed in New Mexico is grazing. Some cotton is also grown. Small but intensive cantalope growing occurs in the Pecos valley, as well as grain and sorghum, which have increased in importance.

The water of the Pecos is high in salinity, with a conductivity in Texas reaching 14,000 μS cm^{-1}. Because of freshwater aquifers discharging into the Pecos, it consists essentially of fresh water near its confluence with the Rio Grande.

Oil and gas fields are common in the Pecos River valley. The flow of the river is greatly influenced by releases due to diversion for irrigation and releases from irrigation. Pollution comes from various sources in the watershed. Nonpoint runoff from croplands, ranges, and pastures is the major mode of pollution entering the Pecos. Feedlots contribute nitrates, ammonia, and organic carbon, which is found in groundwater. Tributaries and runoff from the watershed contribute to an increase in dissolved solids in the river. Farther downstream the inflow of groundwater usually results in a reduction of total dissolved solids. For example, dissolved solids from 1961 to 1968 averaged 7770 mg L^{-1} below Red Bluff, 14,200 mg L^{-1} at Girvin, and 1600 mg L^{-1} at Shumla, thus showing the great effect of groundwater. The groundwater also reduces salinity as one progresses from Orla to Shumla.

In the upper Pecos River above the confluence with West Willow Creek, the water quality is fairly good. As one progresses downstream, the concentration of heavy metals increases. For example, below the reservoir at Santa Rosa, New Mexico, the following substances were found: copper 22 ppm, lead 51 ppm, cadmium 2 ppm, chromium 34 ppm, zinc 49 ppm, nickel 44 ppm, and mercury 0.02 ppm. Often, concentrations are higher in reservoirs than in the free-flowing river.

In the Santa Rosa Reservoir, the fish showed the presence of mercury in their tissues and there seemed to be a strong correlation between mercury concentrations and body weight. The report of the U.S. Fish and Wildlife Service (1992) showed elevated levels of mercury in fish in Santa Rosa Reservoir. In 58% of the fish obtained from Santa Rosa Reservoir, mercury concentrations in the muscle tissue was in excess of 1 ppm dry weight, which is the limit established by EPA. Analyses of sediments above and below the reservoir indicate that the water released from the dam was not having an adverse effect on the river. Upstream from the mines the fish did not show any accumulation of mercury. It appears that mercury and other metals are the result of releases from the mines upstream from the reservoir.

McLemore et al. (1993) concluded that lead and zinc concentrations in the Pecos water samples were above average dissolved concentrations found in stream water, whereas copper was lower. When one compares the concentrations in the Pecos with those in another New Mexico river, the Rio Grande, copper and mercury concentrations were lower and lead concentrations were about the same. Zinc concentrations appeared slightly elevated in the upper Pecos. Comparing Pecos concentrations with New Mexico Water Quality Control Commission District Stream Standards, just below the confluence of Willow Creek was one of two sites with concentrations above the standards for lead and zinc only. Conductivity increased downstream in the Pecos. Mercury and zinc concentrations in sediments were not elevated above what one would expect from common lithologies found in the Pecos drainage (Krauskopf, 1979), with the exceptions of Jacks Creek through just below the village of Pecos. Copper and chromium concentrations in sediments were below crustal abundances except for samples from the Pecos mine water drainage. Mercury, lead, copper, and zinc concentrations in sediments were elevated in samples from above the Pecos mine waste dumps, suggesting that outcroppings, zones of mineralization, and perhaps the outcropping rocks, are sources of contamination as well as waste dumps (McLemore et al., 1993).

Lead concentrations in the Pecos River sediments are above crustal abundance. It is possible that rocks in sediments throughout New Mexico may be enriched in lead relative to crustal abundances. Typically, in the Pecos River basin concentrations in sediments in lakes were higher than those in stream sediments. It is believed that heavy metals in sediments may enter the food chain and eventually concentrate in fish. White and brown precipitates or froth from the seep below the waste dumps were high in zinc and other metals. Fish kills have been reported in the past as a result of contamination by precipitates during high-runoff events. However, U.S. Fish and Wildlife Service reports suggest that the fish kills at Lisbon Spring State Fish Hatchery were due to a combination of water quality factors, including temperature, periodic presence of toxic combination of metals, and possible introduction of disease (*Albuquerque Journal of Fish*, vol. 16, 1993). Fish along the Pecos River between the mine dumps and the factory (a distance of 19 km) appear healthy and show no effects from potential contaminations (McLemore et al., 1993).

In Texas, the largest studies of the chemistry of the water of the Pecos have been between Orla and Shumla. In general, the water quality decreases from the upper river to the lower river. Total dissolved solids can be as high as 20,000 mg L^{-1} during low flows in the lower sections. Water quality is generally improved by contributions of tributaries and flood flows. Irrigation that occurs in the middle and lower reaches combined with oil-field brines from now unused disposal pits near Artesia may add to the total dissolved solids load. The potash industry in the Carlsbad area disposes of a large quantity of salts in unlined lagoons. It is not clear how much the potash waste or oil-field brine contributes to the river's salinity.

Dissolved solids tend to decrease in concentration from Orla to Shell Slab. The largest drop in salinity occurs between Sheffield and Shell Slab. This dilution is believed to be caused by freshwater inflows from sources in the Cretaceous limestone formation of the Edwards–Stockton Plateau, which the river traverses between Sheffield and Shell Slab. Fresh water is also added to the river by Independence Creek, which is located approximately 15 river miles upstream of Shell Slab. The daily average flow of Independence Creek was estimated in 1975–1984 to be 28.4 ft³/s.

General Description of the Biology of the River. The biology of the Pecos River has been determined in New Mexico and Texas. The most extensive studies have been made in Texas. In New Mexico, the insect fauna at Puerto de Luna and Artesia are somewhat different than the distribution at Orla and Shumla in Texas.

Biology of the River in New Mexico

Insects. The taxonomic richness and biomass of insects were greater at Puerto de Luna than at Artesia. The most numerous insects in 1979 that were collected at Puerto de Luna were the Hydropsychidae, *Nectopsyche gracilis*, the chironomid *Chironomus* sp., and the blackflies *Simulium* sp. In general, the insect fauna was much more common at Puerta de Luna than at Artesia. No very large numbers of any one species were found at Artesia. The most common species at Artesia was the dipteran *Calopsectra* sp.

Fish. A study was made of the fish fauna between Lake Sumner and Lake McMillan. In this study *Notropis simus pecosensis* was less abundant and less widely distributed than in the past. In this study this species was found only in the middle section from Fort Sumner to Artesia, although formerly this species was more widely distributed. It is doubtful whether this species can continue to exist between Fort Sumner and Artesia because of periodic dewatered conditions in response to irrigation needs. *N. simus pecosensis* was abundant only between Hagerman and Artesia and in the seepage zone near Fort Sumner.

N. simus is absent between Santa Rosa and Fort Sumner because of the use of fish toxicants and regulation of flow. Other native species have also been eliminated. Since the completion of Sumner Dam in 1937, *N. simus* seems to be declining. It is probable that this species has not been significant in this reach since the completion of McMillan Dam. This species has a prolonged spawning season that seems to end shortly after the first of September. Thirty-one species were recorded in the historic range of this species (Santa Rosa to Fort Sumner). Of these, only 23 were recorded within the present range of this species. Twenty species were collected with *N. simus*. However, only five could be considered frequent associates: *Hybognathus placitus*, *Hybopsis aestivalis*, *Notropis jemezanus*, *N. lutrensis*, and *N. stramineus*. The first three species are ecologically similar to *N. simus*. This species was found in every major habitat encountered from Fort Sumner to Artesia, except in stagnant pools.

Notropis girardi was introduced into the Pecos River. It was probably released as a bait fish below Sumner Dam in 1978. This species dispersed downstream at least 260 km by the autumn of 1981. By 1987, *N. girardi* was verified downstream another 98 km, to the New Mexico–Texas border. Multiple introductions of *N. girardi* may have occurred throughout the Pecos River.

The first collections of *N. girardi* were taken from the Pecos River in New Mexico in 1978 just downstream from Sumner Dam. Collections made from September 1981 to February 1982 indicated that *N. girardi* had become widespread and abundant in the Pecos River. *N. girardi* abundance declined slightly in 1986–1987, and in 1981–1982 and 1986–1987 *N. girardi* was uncommon immediately below Sumner Reservoir and between McMillan and Red Bluff Reservoirs. Relative abundance and population stability varied substantially. The dominant age class for this species was 0 to 1 year, although other classes were found. *N. girardi* is well established in the Pecos River.

Hybognathus placitus, which is an abundant plains streams species, was introduced into the Pecos River. It occurs in the habitat of *N. girardi* in its native range. It may have been introduced accidentally as a bait fish. *H. placitus* is now a most abundant species in the Pecos River, and it comprises more than 80% of the catch at some sites. Its success in the Pecos drainage is probably due to the same factors that have benefited *N. girardi*. The establishment of *H. placitus* coincides with the disappearance of the native *Hybognathus amarus*. This species has not been taken from the Pecos River since 1968.

Pecos Gambusia (*Gambusia nobilis*) once occurred throughout the Pecos drainage. Now it is endangered and found only in seven isolated locations at Bitter Lake National Refuge (Chaves County, New Mexico) and at one locality at Blue Springs (Eddy County, New Mexico). Pecos Gambusia uses submerged cliff overhangs and aquatic vegetation, including *Chara* sp., *Ruppia maritima*, and *Potamogeton pectinatus*, as cover. In Blue Springs, however, Pecos Gambusia uses as habitat *Paspalum distichum*, *Nasturtium* sp., *Chara* sp., and *Eleocharis* sp. No Pecos Gambusia were found where the hardness of the water exceeded 5100 mg L^{-1}. This species prefers to feed at night. Coexisting with Pecos Gambusia are an unidentified species of *Cyprinodon* (Pecos pupfish), *Fundulus zebrinus* (Rio Grande killifish), *Lucania parva* (rainwater killifish), *Notropis lutrensis* (red shiner), and *Dionda episcopa* (roundnosed minnow).

Biology of the River in Texas

Aquatic Vegetation. The aquatic vegetation in two habitats was studied. One is at U.S. Highway 290 southeast of Sheffield at river km 417 in a riffle run on a streambed of calcite and coarse gravel. The periphyton was mainly diatoms, and the area was devoid of rooted vegetation.

The other station in this lower reach of the river was at Shumla in a riffle-run area on gravel and sand substratum. Sparse growth of the green filamentous alga *Zygnema* and the attached diatom *Cymbella* were usually present. In shallow pools and backwaters, dense beds of *Chara* and *Potamogeton* were growing.

A massive fish kill by the golden alga *Prymnesium parvum* occurred in the Pecos River in November 1986. Only six species—*Dionda episcopa*, *Cyprinella lutrensis*, *Cyprinodon* hybrids, *Lucania parva*, *Fundulus grandis*, and *Gambusia affinis*—were able to survive. *Cyprinodon* sp. has been found to be fairly tolerant to the algae toxins.

Invertebrates. A study was made of the invertebrates in the Pecos River northeast of Orla (FM 652), north of Coyanosa (FM 1776), east of Girvin (U.S. 67), and east of Sheffield (U.S. 290), and just upstream of Val Verde County at Shell Slab (Figure 4.2). The fauna was divided into three broad categories: community indicators, prevalence of sensititive taxa, and functional feeding groups. These parameters were used to provide a quantitative means for assessing benthic macroinvertebrate community structure at each of the sampling stations (TNRCC, 1996a).

Macroinvertebrates were found to be quite diverse in the Pecos River. The oligochaetes were most numerous at Orla in the period between 1976 and 1977.

Ostracods were most numerous at Girvin. Macroinvertebrate stream communities were determined in the vicinities of Orla, Girvin, Sheffield, and Shumla. Seventeen species were identified at Orla, 21 at Girvin, 24 at Sheffield, and 50 at Shumla. This, no doubt, is because of the better quality of the water for diverse invertebrates at Shumla, where the salinity was much less.

Orla, Texas Just Upstream of the FM 652 Bridge.   The benthic macroinvertebrate community here was limited in taxa due to the poor habitats and relatively high salinity. The number of invertebrate genera was only eight, and this site had the fewest number of individuals captured.

North of Coyanosa (FM 1776).   The flow is much less at this station than at Orla. This is caused by extensive withdrawals from the river via irrigation canals between Mentone and this site. Also, there is loss in flow at this station, due to evaporation and transpiration. The macroinvertebrate community at Coyanosa indicated saline conditions. Coyanosa was the second-most-saline site, after Girvin. Although salinity is higher than at Orla, the habitats were so much better that the site supported a greater number of benthic macroinvertebrates. Three hundred and twenty-seven individuals were collected, which is the second highest number collected at any site. Sixty-three percent were the amphipod, *Hyalella azteca*. Other taxa present in abundance were *Berosus* sp., *Argia* sp., four species of chironomids, two Trichoptera genera, and one species of the Anisoptera. All of these species are relatively low in pollution tolerance. Most of the species in this community were gatherers, including an abundant population of *Hyalella azteca*. Decaying vegetation formed a plentiful source of food. Predators were well represented. A total of 327 macroinvertebrates were collected, representing 12 genera. The largest population was amphipods.

East of Girvin on U.S. 67.   At Girvin, because of the high salinity value of the water, the taxa were salt tolerant. *Hyalella azteca* was the dominant species, comprising 53.9% of the community. Other abundant species were the beetle *Berosus* sp. (69 specimens), the damselfly *Argia* sp. (17 specimens), and the mayfly *Callibaetis* sp. (1 specimen). The gastropod *Physella virgata* was found in high numbers (30 individuals). The community trophic structure was dominated by gatherers, primarily *Hyalella azteca* (an amphipod) and *Berosus* sp. (a beetle). These two comprised 75% of the community.

East of Sheffield at U.S. 290.   In this region the river flows into a geological area characterized by Cretaceous limestone formations. Riffles were more common here than upstream, and fairly deep pools were present that attracted fish (TNRCC, 1996b).

At Sheffield there was an inflow of fresh water from Cretaceous limestone formations. Only 55 invertebrates were captured. Most of these were the gastropod (snail) *Physella virgata*, which comprised 49.1% of the population. Eleven genera were collected, six were chironomids of varying pollution tolerance (Table 4.7). Similar to the three saline sites upstream, *Hyalella azteca* and *Berosus* sp. were present. The community structure was characterized by collector-gatherers, including *Hyalella azteca*, some coleopterans (beetles), a few chironomids, and scrapers dominated by *Physella virgata*. These scrapers and gatherers comprised about 90% of the community, indicating that primary food sources were decaying vegetation and periphytic growth on the substrate. The other most common group was the true flies (TNRCC, 1996a).

Shell Slab. The collecting area at Shell Slab consisted of a much better habitat for aquatic life. This was caused by increased flow and the predominance of gravel, cobble, and boulder substrates. The convergence of Independence Creek approximately 15 mi upstream also increased the number of species available for this habitat. Twenty-eight genera were collected, and more individuals were captured than at any other site. The dominant genus and species was *Macrelmis texana*, which comprised 27.7% of the community. The numbers of scrapers and filter feeders were approximately the same, reflecting a balanced food source of periphyton and suspended organic matter. The site exhibited a low biotic index, indicating that the benthic macroinvertebrate community as a whole is relatively pollution intolerant. Ten EPT (Ephemeroptera, Plecoptera, Trichoptera) taxa were collected, five times more than at the site with the next-highest number, which was Coyanosa. Diversity was high in this area.

SUMMARY—INVERTEBRATES

In summary, the total oligochaetes were much more numerous at Orla than at the other stations listed above, whereas ostracods were much more numerous at Girvin. Ephemeroptera were much more common at Shumla than at the other stations. Elmid beetles were also much more common at Shumla, whereas unidentified species of *Berosus* were more common at Girvin. *Cricotopus bicinctus* was most common at Sheffield and *Pseudochironomus* was much more common at Sheffield, while *Tanytarsus* was much more common at Girvin followed by Orla. The greatest number of species were taken at Shumla, followed by Sheffield, with fewer species at Orla. Taxonomic richness was similar at Orla, Girvin, and Sheffield but was almost double at Shumla.

Taxonomic richness of macroinvertebrates increased as one proceeded downstream (17 species at Orla, 21 at Girvin, 26 at Sheffield, and 50 at Shumla). These differences are probably caused by water quality, primarily salinity and total dissolved solids. A similar longitudinal pattern of diversity was evident. Regulation of flow at Red Bluff Reservoir probably represented the greatest variation in occurrence, affecting invertebrates at Orla.

Taxonomic richness (number of species in a given number of individuals) was similar at Orla, Girvin, and Sheffield but was almost double at Shumla. Variations in occurrence of invertebrates at Orla seemed to be associated with flow regulations at Red Bluff. Variation in occurrence at Girvin was similar in autumn and winter for each year, but spring and summer values varied from year to year. Variation appears to depend on autotrophic standing crops, which were roughly associated with seasonality.

At Orla, Girvin, and Sheffield, the seasonal diversity of insects seemed to be caused more by the dominance or shift in dominance of one or more taxa than by a change in the size of populations of most of the species. Conversely, seasonal changes in diversity at Shumla were induced to a greater degree by changes in the number of taxa than by changes in dominance, as *d* varied directly with the number of taxa, and *d* and *r* were more independent.

Seasonal diversity at all stations had no direct correlation to seasonal physico-chemical changes, which varied irregularly. At Orla the diversity of species seemed to vary directly with the duration of constant-volume releases from Red Bluff Reservoir prior to sampling. The insect fauna of the Pecos River is diverse but apparently has not been as well studied as the fish fauna.

Over time, the great shifts in climate, particularly toward aridity during the late Tertiary, has resulted in a considerable extinction of many species of the fauna. In Texas, the changes in salinity and other water characteristics have greatly influenced or caused the changes in aquatic life. This is shown clearly by the differences in fauna and flora at Orla and Girvin as compared to Sheffield. The reduction in salinity, particularly at Shumla, shows the importance of this river characteristic on the fauna of the river.

In conclusion, it is quite evident that the chemical characteristics of the water, together with the influx of water from irrigation and also the influx of groundwater, has had a profound effect on the structure of the aquatic communities in various reaches of the Pecos River.

SUMMARY—FISH

The reach of the Columbia between Malaga, New Mexico, and Sheffield, Texas, consists of fresh, somewhat brackish water. Present in these waters were the omnivorous *Dorosoma cepedianum*, *Cyprinella (= Notropis) proserpina* an omnivore, and *Cyprinus carpio* an omnivore. Also present was the herbivore *Dionda espiscopa* and *Astyanax mexicanus*, which is a carnivore-omnivore. It was typically found among gravel and rocks. Also present was the omnivore *Cyprinodon pecosensis*, which was in faster-flowing water among gravel and rocks. In this habitat also were the omnivore *Cyprinodon* (a hybrid) and *Lucania parva*. The *Cyprinodons* are probably omnivores. Among the vegetation and also in areas where the water was relatively slow was present *Gambusia affinis*, an omnivore. The carnivore *Menidia beryllina* was found in pools but also in faster-flowing water. Also in the brackish reach between Malaga, New Mexico, and Sheffield, Texas, was found *Lepomis cyanellus*, which is a carnivore. It was found both in pools and in faster-flowing water. *Lepomis megalotis*, a carnivore, was found in this locality among vegetation and in pools.

In the lower Pecos River, between Sheffield, Texas, and the confluence of the Rio Grande, was *Dorosoma cepedianum*, and omnivore; *Carpiodes carpio*, a carnivore; and *Moxostoma congestum*, a carnivore. They were mainly found in muddy pools. *Cyprinella (= Notropis)* proserpina, an omnivore, was also found in this lower part of the Pecos River along with *Cyprinella (= Notropis)* venusta, a carnivore, which was found in pools and slower-flowing water as was the omnivorous *Cyprinus carpio*, which was in among vegetation. The omnivorous *Hybopsis aestivalis* was found in pool in faster-flowing water as was *Notropis amabilis*. Other species found in the lower Pecos River was *Notropis braytoni*, the omnivorous *Notropis jemezanus*, and the omnivorous *Pimephales promelas* and *P. vigilax*. The corixid *Astyanax mexicanus*, which is a carnivore-omnivore living typically in faster water, was also found here. Another species found in this habitat was the catfish *Ictalurus lupus*, which was typically found in pools and is an omnivore as was *I. punctatus*, and *Pylodictus olivaris*, which is a carnivore. Among gravel and rocks were the omnivorous *Cyprinodon pecosensis* and *Cyprinodon* hybrids of various types, which are typically found in faster flowing water. In pools were *Fundulus grandis*, an omnivore, and *Lucania parva*, which was found mainly in and among vegetation. Present also in this reach of the river were *Gambusia affinis*, an omnivore, and *Menidia beryllina*, a carnivore. Other species found in the lower reach of the Pecos River between Sheffield and the junction of the Pecos with the Rio Grande were *Lepomis cyanellus*, a carnivore, which was mainly in pools but also in faster-flowing water. *Lepomis megalotis* was found in among the vegetation and in pools. It is a carnivore. *Micropterus salmonides*, a carnivore, was found among the vegetation, but also in faster-flowing water. Other species found in the habitat were *Cichlasoma cyanoguttatum* and *Etheostoma grahami*, which are carnivores.

It is interesting to note, that in this lower reach of the Pecos River, the fish seem to be more characteristic of fresh to relatively high conductivity water. The fauna are quite diverse and consist of omnivores and carnivores.

BIBLIOGRAPHY

Bailey, L. 1974. Water quality segment report for segment 2311. Pecos River Rep. WQS-8. Texas Water Quality Board, Austin, Texas.

Bednarz, J. C. 1979. Ecology and status of the Pecos Gambusia, *Gambusia nobilis* (Poeciliidae), in New Mexico. Southwest Nat. 24(2): 311–322.

Bestgen, K. R., and S. P. Platania. 1990. Extirpation of *Notropis simus simus* (Cope) and *Notropis orca* (Woolman) (Pisces: Cyprinidae) from the Rio Grande in New Mexico. Occas. Pap. Mus. Southwest. Biol. 6: 1–8.

Bestgen, K. R., and S. P. Platania. 1991. Status and conservation of the Rio Grande silvery minnow, *Hybognathus amarus*. Southwest. Nat. 36(2): 225–232.

Bestgen, K. R., S. P. Platania, J. E. Brooks, and D. L. Propst. 1989. Dispersal and life history of traits of *Notropis girardi* (Cypriniformes: Cyprinidae) introduced into the Pecos River, New Mexico. Am. Midl. Nat. 122: 228–235.

Bogener, S. D. 1997. Ditches across the desert: a story of irrigation along New Mexico's Pecos River. Ph.D. thesis. Texas Tech. Univ., Lubbock, Texas.

Brandvold, L. A., and D. K., Brandvold. 1990. A compilation of trace-metal values in water and sediments collected along the Rio Grande and its tributaries in New Mexico, data from selected published and unpublished sources. Open-file Rpt. 359. New Mexico Bureau of Mines and Mineral Resources, Santa Fe, N. Mex. 103 pp.

Brandvold, D. K., C. J. Popp, and L. A. Brandvold. 1981. Transport mechanisms in sediment rich streams, heavy metal and nutrient load of the Rio San Jose–Rio Puerco systems. Rep. 132. New Mexico Water Resources Research Institute, Las Cruces, N. Mex.

Buckner, H. D., E. R. Carrillo, H. J. Davidson, and W. J. Shelby. 1989. Water resources data. Water-Data Rep. TX-88-3. Texas water year 1988. U.S. Geological Survey, Washington, D.C.

Claassen, H. C. 1981. Estimation of calcium sulfate solution rate and effective aquifer surface area in a ground-water system near Carlsbad, New Mexico. Groundwater 19(3): 287–297.

Crowley, D. E., and J. E. Sublette. 1987a. Distribution of fishes in the Black River Drainage, Eddy County, New Mexico. Southwest. Nat. 32(2): 213–221.

Crowley, D. E., and J. E. Sublette. 1987b. Food habits of *Moxostoma congestum* and *Cycleptus elongatus* (Catostomidae: Cypriniformes) in Black River, Eddy County, New Mexico. Southwest. Nat. 32(3): 411–413.

Davis, J. R. 1980. Species composition and diversity of benthic macroinvertebrate populations of the Pecos River, Texas. Southwest. Nat. 25(2): 241–256.

Davis, J. R. 1981. Diet of the Pecos River pupfish, *Cyprinodon pecosensis* (Cyprinodontidae). Southwest. Nat. 25(4): 535–540.

Davis, J. R. 1987. Faunal characteristics of a saline stream in the northern Chihuahuan Desert. Contributed papers of the 2nd Symposium on Resources of the Chihuahuan Desert region: United States and Mexico. Chihuahuan Desert Research Institute, Alpine, Texas. 18 pp.

Echelle, A. A. 1975. A multivariate analysis of variation in an endangered fish, *Cyprinodon elegans*, with an assessment of population status. TX J. Sci. 26: 529–538.

Echelle, A. A., and P. J. Conner. 1989. Rapid, geographically extensive genetic introgression after secondary contact between two pupfish species (Cyprinodon: Cyprinodontidae). Evolution 43(4): 717–727.

Echelle, A. A., and A. F. Echelle. 1978. The Pecos River pupfish, *Cyprinodon pecosensis* n. sp. (Cyprinodontidae), with comments on its evolutionary origin. Copeia 1978: 569–582.

Echelle, A. F., and A. A. Echelle. 1986. Geographic variation in morphology of a spring-dwelling desert fish, *Gambusia nobilis* (Poeciliidae). Southwest. Nat. 31(4): 459–468.

Gibbons, W. J. 2001. Pecos river data. U.S. Geological Survey, Washington, D.C.

Grozier, R. U., H. W. Albert, J. F. Blakey, and C. H. Hembree. 1966. Water-delivery and low-flow studies Pecos River, Texas: quantity and quality, 1964 and 1965. Texas Water Dev. Board Rep. 22. Prepared by the U.S. Geological Survey in cooperation with the Texas Water Development Board and the Red Bluff Water Power Control District.

Grozier, R. U., H. R. Hejl, and C. H. Hembree. 1968. Water delivery study: Pecos River, Texas, quantity and quality, 1967. Report 76. Texas Water Development Board, Austin, Texas. 16 pp.

Harman, W. J., M. S. Loden, and J. R. Davis. 1979. Aquatic Oligochaeta new to North America with some further records of species from Texas. Southwest. Nat. 24: 509–525.

Harrell, H. L. 1978. Response of the Devil's River (Texas) fish community to flooding. Copeia 1978: 60–68.

Hatch, M. D., W. H. Baltosser, and C. G. Smitt. 1985. Life history and ecology of the bluntnose shiner (*Notropis simus pecosensis*) in the Pecos River of New Mexico. Southwest. Nat. 30: 555–562.

Hawman, F. 1995. Heavy metal concentrations in sediment and fish in Santa Rosa Reservoir and the Pecos River, New Mexico. M.S. Thesis. New Mexico Institute of Mining and Technology, Socorro, N. Mex.

Hernandez, J. W. 1971. Management alternatives in the use of the water resources of Pecos River basin in New Mexico. WRRI Rep. 12. New Mexico Water Resources Research Institute, Las Cruces, N. Mex.

Hernandez, J. W., and T. J. Eaton, Jr. 1988. A bibliography pertaining to the Pecos River basin in New Mexico. Water Resources Research Institute in cooperation with Engineering Experimental Station, New Mexico State Univ., University Park, N. Mex.

Hoagstrom, C. 1994. Relative abundance and niche partitioning in the fish community of the Pecos River, Pecos County, Texas. M.S. thesis. Sul Ross State Univ., Alpine, Texas.

Hubbs, C. L., and R. R. Miller. 1978. *Notropis panarcys* n. sp. and *N. proserpinus*, cyprinid fishes of subgenus *Cyprinella*, each inhabiting a discrete section of the Rio Grande complex. Copeia 1978(4): 582–592.

Hunt, A. P. 1997. Proc. 42nd Annual New Mexican Water Conference. Water issues of eastern New Mexico: an overview of the geology of northeastern New Mexico.

Itzkowitz, M. 1969. Observations on the breeding behavior of *Cyprinodon elegans* in swift water. TX J. Sci. 21: 229–231.

Jacobi, G. Z., and L. R. Smolka. 1982. Upper Pecos River water quality study, 1980–81. Water Pollution Control Bureau, Environmental Improvement Division, New Mexico Health and Environment Department, Santa Fe, N. Mex.

Katz, S. R., and P. Katz. 1985. The pre-history of the Carlsbad basin SE New Mexico. Bureau of Reclamation Contract 3-CS-57-01690.

Kelsh, S. W., and F. S. Hendricks. 1990. Distribution of the headwater catfish *Ictalurus lupus* (Osteichthyes: Ictaluridae). Southwest. Nat. 35(3): 292–297.

Koster, W. J. 1957. Guide to the fishes of New Mexico. Univ. New Mexico Press, Albuquerque. N. Mex. 116 pp.

Krauskopf, F. K. B. 1979. Introduction to geochemistry. McGraw-Hill Book Co., New York, New York. 617 pp.

Larkin, T. J., and G. W. Bomar. 1983. Climatic atlas of Texas. Report LP-192. Texas Department of Water Resources, Austin, Texas. 152 pp.

Liu, R. K. 1969. The comparative behavior of allopatric species (Teleostei-Cyprinodontidae: *Cyprinodon*). Ph.D. thesis, University of California, Los Angeles.

Loomis, G. L. 1966. The assault on salt in the Pecos. Reclam. Era 52(2): 37–39.

Looper, T. 1995. Trace metal analysis of Pecos River water, sediment and suspended sediment. M.S. thesis. New Mexico Institute of Mining and Technology, Socorro, N. Mex.

Lynch, T. R., C. J. Popp, and G. Z. Jacobi. 1988. Aquatic insects as environmental monitors of trace metal contamination: Red River, New Mexico. Water, Air, Soil Pollut. 42: 19–31.

McAda, D. P., and T. D. Morrison. 1993. Sources of information and data pertaining to geohydrology in the vicinity of the Roswell basin in parts of Chaves, Eddy, DeBaca, Guadalupe, Lincoln, and Otero counties, New Mexico. Open-File Rep. 93–144. U.S. Geological Survey, Washington, D.C.

McLemore, V. T., L. A. Brandvold, and D. K. Brandvold. 1993. A reconnaissance study of mercury and base metal concentrations in water, and stream- and lake-sediment samples along the Pecos River, eastern New Mexico. Pp. 339–351. *In:* New Mexico Geological Society guidebook, 44th field conf., Carlsbad Region, New Mexico and West Texas.

Mendieta, H. B. 1974. Reconnaissance of the chemical quality of surface waters of the Rio Grande basin, Texas. Rep. 80. Texas Water Development Board, Austin, Texas. 109 pp.

New Mexico Water Quality Control Commission. 1976. Pecos River Basin Plan. NMWQCC. Santa Fe, N. Mex.

O'Leary, M. C. 1980. Texas v. New Mexico: the Pecos River compact litigation. Nat. Resour. J. 20: 395–410.

Pease, T. C., V. T. McLemore, L. A. Brandvold, and A. M. Hossain. 1995. The effect of particle-size distribution on the base-metal geochemistry of stream sediments from the upper Pecos River, New Mexico. N. Mex. Geol., May, pp. 27–28.

Pennak, R. W. 1978. Fresh-water invertebrates of the United States, 2nd ed. Wiley, New York. 803 pp.

Quinlan, P. T. 1982. Climatic change and water availability in the Rio Grande and Pecos River basins. M.S. Thesis. Univ. Arizona, Tucson, Ariz.

Reuther, R. 1994. Mercury accumulation in sediment and fish from rivers affected by alluvial gold mining in the Brazilian Madeira River basin, Amazon. Environ. Monitor. Assess. 32: 239–258.

Rhodes, K., and C. Hubbs. 1992. Recovery of Pecos River fishes from a red tide fish kill. Southwest. Nat. 37(2): 178–187.

Rothfork, J. 1997. Heavy metal resistant algae in the Pecos. M.S. thesis. New Mexico Institute of Mining and Technology, Socorro, N. Mex.

Smith, M. L., and R. R. Miller. 1986. The evolution of the Rio Grande basin as inferred from its fish fauna. Pp. 457–485. *In:* C. H. Hocutt and E. O. Wiley (eds.)., The zoogeography of North American freshwater fishes. Wiley, New York. 866 pp.

Smolka, L. R., and D. F. Tague. 1984. Intensive survey of lower Pecos River in vicinity of Calsbad, New Mexico. EID/SWQ-84/8. New Mexico Environmental Improvement Division, Santa Fe, N. Mex.

Spiers, V. L., and H. R. Hejl, Jr. 1970. Quantity and quality of low flow in the Pecos River below Girvin, Texas, February 6–9, 1968. Rep. 107, Texas Water Development Board, Austin, Texas.

Stevenson, M. M., and T. M. Buchanan. 1973. An analysis of hybridization between the cyprinodont fishes *Cyprinodon variegates* and *C. elegans*. Copeia 1973: 682–692.

Sublette, J. E., M. D. Hatch, and M. Sublette. 1990. The fishes of New Mexico. Univ. New Mexico Press, Albuquerque, N. Mex.

Taylor, D. W. 1985. *Pecosorbis*, a new genus of freshwater snails (Planorbidae) from New Mexico. Circ. 194. New Mexico Bureau of Mines and Mineral Resources, Santa Fe, N. Mex.

Texas Natural Resources Conservation Commission. 1996a. Benthic macroinvertebrate community struc-
    ture in relation to water quality and habitat in the upper Pecos River, Texas. AS-107/SR. Prepared by
    Greg L. Larson, Field Operations Division. TNRCC, Austin, Texas. 29 pp.
Texas Natural Resources Conservation Commission. 1996b. Fish community structure in relation to water
    quality and habitat of the upper Pecos River, Texas. AS-095/SR. Prepared by Greg L. Larson, Field
    Operations Division. TNRCC, Austin, Texas. 29 pp.
Texas Water Quality Board. 1974. Segment No. 2311 Pecos River. Rep. WQS8. TWQB, Austin, Texas.
U.S. Fish and Wildlife Service. 1992. Pecos bluntnose shiner recovery plan. U.S. fish and Wildlife Service,
    Region 2, Albuquerque, N. Mex. 57 pp.
U.S. Geological Survey. 1970–1973. Water resources data for Texas, Parts 1 and 2. USGS, Washington,
    D.C.
U.S. Geological Survey. 1975. Proc. 3rd National Conference, Water Resources Division, Sept. 28–Oct.
    3. Albuquerque, N. Mex.
U.S. Geological Survey. U.S. Bureau of Mines, and N.M. Bureau of Mines and Mineral Resources. 1980.
    Mineral resources of the Pecos wilderness and adjacent areas, Santa Fe, San Miguel, Mora, Rio Arriba,
    and Taos counties, N.M. Open-file Rep. 80-382. U.S. Geological Survey, Washington, D.C. 103 pp.
Wetzel, R. G. 1975. Limnology. W. B. Saunders Co., Philadelphia. 743 pp.
White, D. H., and A. J. Krynitsky. 1986. Wildlife in some areas of New Mexico and Texas accumulate el-
    evated DDE residues, 1983. Arch. Environ. Contam. Toxicol. 15: 149–157.
Wilhm, J. L. 1970. Range of diversity index in benthic macroinvertebrate populations. J. Water Pollut.
    Control. Fed. 42: 221–224.

Index

A

Abiquiu Dam, 128
Ablabesmyia:
  Pecos riverine system, 210, 223, 224
  Rio Grande, 132
  Sabine River, 24, 26, 27
Ablabesmyia rhamphe, 24, 26, 27
Acanthocephala, 205
Acequia Madre Canal, 119
Achnanthes, 28, 58
Achnanthes affinis, 75, 96, 98
Achnanthes exigua, 14, 25, 27, 75, 96
Achnanthes minutissima, 25, 27
Achnanthes submontana, 25
Achnanthes temperei, 56
Acidity, Sabine River, 6
Acme-Bitter Lakes, 192
Acorus calamus, 70, 75, 96
Acroneuria abnormis, 223, 224
Actinophrys sol, 97
Agmenellum quadruplicatum, 25
Agonostomus monticola, 145, 147, 148
Agosia chrysogaster, 139
Alamogordo Creek, 174

Algae:
  Guadalupe River, 69, 70, 74, 75, 96
  Pecos riverine system, 196, 205, 206, 211, 222
  Rio Grande, 131, 138
  Sabine River, 9, 14, 25
  (See also specific types)
Algal-grazing mayflies, Guadalupe River, 74
Alkalinity:
  Guadalupe River, 66
  Pecos riverine system, 185, 191, 192
  Rio Grande, 120
  Sabine River, 6
Alligator gar, 10, 142
Aluminum, Pecos riverine system, 191
Amaranthus tamariscina, 14, 24, 25
Amazon molly, 144
Amblema perplicata, 72
Amblema plicata, 26, 27
Amblema plicata var. *perplicata,* 96–98
Ambloplites rupestris, 73, 142
Ambrysus, 159–161, 202, 223, 224
Ambrysus mormona, 222–224